Insulating Materials for Future Power Systems: Performance Analysis, Defect Detection and Condition Assessment

Insulating Materials for Future Power Systems: Performance Analysis, Defect Detection and Condition Assessment

Guest Editors

Xianhao Fan
Chuanyang Li
Fangwei Liang
Jiefeng Liu
Weidong Liu

MDPI

Basel • Beijing • Wuhan • Barcelona • Belgrade • Novi Sad • Cluj • Manchester

Guest Editors

Xianhao Fan
Department of Electrical Engineering
Tsinghua University
Beijing
China

Chuanyang Li
Department of Electrical Engineering
Tsinghua University
Beijing
China

Fangwei Liang
Department of Electrical Engineering
Tsinghua University
Beijing
China

Jiefeng Liu
Department of Electrical Engineering
Guangxi University
Nanning
China

Weidong Liu
Department of Electrical Engineering
Tsinghua University
Beijing
China

Editorial Office
MDPI AG
Grosspeteranlage 5
4052 Basel, Switzerland

This is a reprint of the Special Issue, published open access by the journal *Energies* (ISSN 1996-1073), freely accessible at: https://www.mdpi.com/journal/energies/special_issues/G129Z132L1.

For citation purposes, cite each article independently as indicated on the article page online and as indicated below:

Lastname, A.A.; Lastname, B.B. Article Title. *Journal Name* **Year**, *Volume Number*, Page Range.

ISBN 978-3-7258-5203-1 (Hbk)
ISBN 978-3-7258-5204-8 (PDF)
https://doi.org/10.3390/books978-3-7258-5204-8

Contents

About the Editors

Xianhao Fan

Xianhao Fan, Tsinghua University. His current research interests include intelligent optimization, machine learning, and condition assessment for power equipment. He received his B.Sc. degree in electrical engineering in 2018 from Guangxi University, where he subsequently earned his Ph.D. degree in 2023. He is currently conducting postdoctoral research at Tsinghua University. He is the author and co-author of over 60 SCI papers. He served as a Guest Editor of *High Voltage*, *Energies*, and *Front. Mater.*, reviewer for *IEEE TNNLS*, *IEEE TTE*, *IEEE TIM*, *IEEE TDEI*, *IEEE TPD*, *High Volt.*, *ENG Appl. Artif. Intel*, *Neural Networks*, etc.

Chuanyang Li

Chuanyang Li received his Ph.D. degree in Electrical Engineering from Tsinghua University, Beijing, China, in 2018. From 2018 to 2022, he held postdoctoral positions at the University of Bologna, Italy; the University of Connecticut, USA; and The Chinese University of Hong Kong. Since June 2022, he has been at Tsinghua University, where he is currently an Associate Researcher, focusing on GIS/GIL fundamentals, key technologies, and equipment development.

Fangwei Liang

Fangwei Liang, Tsinghua University. His current research interests include space charge and condition assessment for power equipment. He received his B.Sc. degree from the North University of China, Taiyuan, China, in 2013. He received a Ph.D. degree in Electrical Engineering from Hunan University in 2022. He is also engaged in postdoctoral research in State Key Laboratory of Power Systems, Tsinghua University.

Jiefeng Liu

Jiefeng Liu was born in Hebei, China, in 1985. He received his M.S. and Ph.D. degrees in Electrical Engineering from Chongqing University, Chongqing, China, in 2011 and 2015, respectively. In 2018, he joined Guangxi University, Nanning, China, where he is an Associate Professor at the School of Electrical Engineering. Throughout his carreer, he has authored and co-authored over 90 articles published in journals and presented at conferences. His research interests include insulation condition assessments and fault diagnosis for transformers.

Weidong Liu

Weidong Liu received his B.Sc. degree from the Shanghai Jiaotong University, Shanghai, China, in 1982, and M.Sc. and Ph.D. degrees from Tsinghua University, Beijing, China, in 1985 and 1989, respectively. From 1990–1992, he was a Visiting Scholar with the University of Liverpool, Liverpool, U.K, and is currently a Professor with the Department of Electrical Engineering at Tsinghua University. His research interests include high-voltage switch equipment, online detection, and fault diagnosis of high-voltage equipment.

energies

MDPI

Article

Comprehensive Properties of Grafted Polypropylene Insulation Materials for AC/DC Distribution Power Cables

Shangshi Huang [1,*], Yuxiao Zhou [2], Shixun Hu [1], Hao Yuan [3], Jun Yuan [2], Changlong Yang [2], Jun Hu [1], Qi Li [1] and Jinliang He [1]

1 Department of Electrical Engineering, Tsinghua University, Beijing 100084, China; rugals@yeah.net (S.H.); hjun@tsinghua.edu.cn (J.H.); qili1020@tsinghua.edu.cn (Q.L.); hejl@tsinghua.edu.cn (J.H.)

2 State Grid Liaoning Electric Power Co., Ltd., Shenyang 110004, China; 15140080905@163.com (Y.Z.); 13701161995@163.com (J.Y.); yangcl_sy@ln.sgcc.com.cn (C.Y.)

3 Sinopec (Beijing) Chemical Industry Research Institute Co., Ltd., Beijing 100013, China; yuanhao.bjhy@sinopec.com

* Correspondence: huangshangshi@mail.tsinghua.edu.cn

Abstract: Polypropylene (PP) exhibits excellent insulation properties, high thermo-stability, and recyclable nature, thus expected to be the next-generation insulation material for HV cable application. Chemical grafting modification is an effective technology to improve the electrical properties of polypropylene. In this paper, we develop and report a new-type grafted PP insulation material by water-phase grafting technology. The comprehensive properties including electrical, thermal, and mechanical of it are tested and compared with traditional cable insulation material—crosslinked polyethylene (XLPE). The results show that the grafted PP holds superior thermal properties and enough mechanical flexibility. The electrical properties are of significant advantages, including resistivity enhanced by nearly two degrees of magnitudes, AC/DC breakdown strength raised by over 20%, and obviously suppressed space charge accumulation. These results indicate that grafted PP is very suitable for application in HV cable systems, either AC or DC. This research lays a foundation for the research and development of the next-generation recyclable polypropylene cables at higher voltage levels.

Keywords: grafted polypropylene; recyclable insulation materials; properties

1. Introduction

Power cables will become mainstream in urban power transmission and distribution, long-distance trans-sea power transmission, far offshore wind power, and onshore new energy power transmission. With the low-carbon and environmental protection development of power equipment, it is urgent to develop cable insulation materials with recyclable nature, high economical efficiency, and high thermostability [1,2]. At present, cross-linked polyethylene(XLPE) is widely used as power cable insulation material, but XLPE cannot be recycled after the end of the cable lifetime, and the manufacturing processes of XLPE cables include cross-linking and degassing, which increases energy consumption and carbon emissions. Therefore, in order to reduce the carbon emission of the whole life cycle and realize recycling after the end of cable lifetime, polypropylene (PP) has drawn great attention from both academia and industry because of its superior electrical properties, high thermostability, and recyclable nature, as shown in below Table 1 [3–5].

The modification method of PP mainly consists of blending, copolymerization, nanodoping, and grafting modification. The blending is the PP matrix mixing with other polymer matrices. The copolymerization modification is propylene monomer copolymerizing with other olefin monomers. The most commonly used blend/copolymerization routes of PP insulation include polypropylene/elastomer blends [6], ethylene-propylene copolymers [7], ethylene-octene copolymers [8], which can reduce the elastic modulus of PP, whereas also the electrical insulation properties. The nanodoping refers to doping a certain amount of nanoparticles into a PP matrix, such as MgO [5], ZnO [9], SiO_2 [10], which can significantly improve the electrical insulation properties, but the agglomeration of nanoparticles in PP matrix cannot be completely avoided, it restricts the feasibility of its large-scale production [11]. The grafting modification is to graft the chemical groups onto the molecular chains of PP, which is a modification method at a microscopic level to realize the improvement of macroscopic electrical properties. The grafted groups can introduce a deep trap into the material, thus improving the electrical insulation properties [12]. Furthermore, compared with nanodoping, grafting modification has no agglomeration, which is more suitable for large-scale industrial preparation and production.

Table 1. Comprehensive properties of PP versus XLPE [5].

Properties	PP	XLPE
DC volume resistivity (Ω·m)	1.7×10^{15}	0.9×10^{15}
DC breakdown strength (MV/m)	399	300
Operation temperature (°C)	Over 90	70–90
Thermal nature	Thermoplastic	Thermoset
Mechanical property	Hard and brittle	Soft and flexible

To evaluate the comprehensive performances of grafting modification, this paper carries out an experimental study on the comprehensive properties of grafted polypropylene insulation material for AC/DC distribution power cables. The AC and DC electrical, thermal, and mechanical properties of grafted PP material were tested and analyzed. We also selected two mature XLPE cable insulation materials as references to make a direct comparison. This work reveals the pros and cons of PP-based and XLPE insulation materials, and it would provide a valuable reference for the research and development of next-generation recyclable polypro-pylene cables at higher voltage levels.

2. Materials Preparation and Characterization

2.1. Materials Preparation

The styrene grafted polypropylene was marked as PPg, which was jointly developed by Tsinghua University and Sinopec (Beijing) Chemical Industry Research Institute. The PP matrix material used for PPg preparation was from Sinopec, and the brand name was NS20. The molecular weight is about 370,000 g/mol. Water-phase grafting technology was adopted for the preparation. Benzoyl peroxide (BPO) was used as the initiator in the reaction process. The free radical formed by heat decomposition of the initiator, had taken the hydrogen atoms of PP to form PP macromolecular radical. The PP macromolecular radical resulted in a grafting reaction with the styrene monomer, as shown in Figure 1 [11]. The reaction of this technology is mild and easy to control, simple operation, of few residues, high purity, and has the feasibility of large-scale preparation and industrial production. For comparison, two commercial XLPE cable insulation materials in distribution lines were selected and produced by Zhejiang Wanma Macromolecule Material Group Co., Ltd. (Hangzhou, China) and Shanghai New Shanghua Polymer Material Co., Ltd. (Shanghai, China), the card number YJ-35 and 4205EC-35 marked as XLPE1 and XLPE2 respectively.

By plate vulcanizer pressing, the three materials PPg, XLPE1, and XLPE2 were prepared into film samples with the required thickness for each test. The XLPEs were first pressed for 10 min at 120 °C and 15 MPa, then crosslinked for 15 min at 180 °C, during which the polyethylene can chemically crosslink into a polymeric network. Finally, it was cooled to room temperature through circulating water. The PPg was pressed for 10 min at 200 °C and 15 MPa, finally cooled to room temperature in the same manner. The prepared samples were left in a vacuum oven for 24 h at 70 °C.

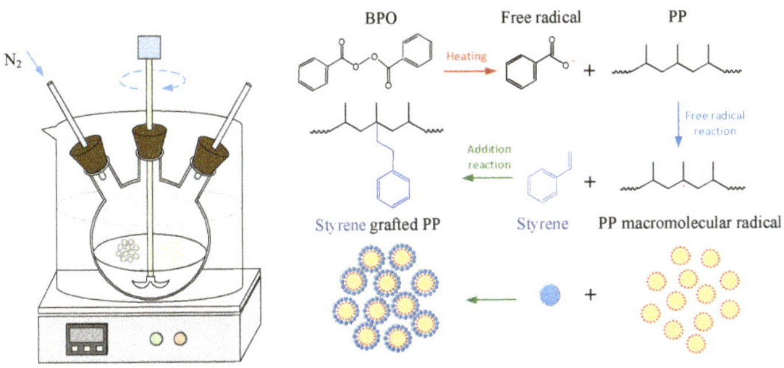

Figure 1. The reaction of styrene grafted PP.

2.2. Materials Characterization

The Fourier transform infrared spectrometer (FTIR) was used for chemical group characterization. Transmission mode was used with the scanning number of 32 and resolution of 4 cm^{-1}, and scanning range from 400 cm^{-1} to 4000 cm^{-1}. To eliminate the interference of unreacted and self-polymerized grafting monomer, the Soxhlet extraction was adopted to remove the ungrafted styrene. The prepared grafted PP was extracted by ethyl acetate for 24 h, then dried and prepared into a film sample as described in Section 2.1. Then the FTIR absorption spectra of the purified samples were tested. Besides, to identify the characteristic peaks of grafted PP, pure PP without any modification is also prepared into a film sample and the FTIR is tested [13].

The Differential scanning calorimeter (DSC) was used for heat flow characterization. The samples were heated to 200 °C and maintained for 5 min, then cooled to 30 °C at the rate of 10 °C/min for testing the hot flow in the crystallization process. And the samples were re-heated to 200 °C at the rate of 10 °C/min for testing in the melting process.

The universal tension tester was used to test the tensile properties. The samples were dumbbell shaped with an average thickness of 200 μm, parallel part length of 33 mm, narrow width of 6 mm, and a tensile rate of 20 mm/min. Each kind of material was tested 5 times, and the measurement results were the mean and standard deviation values.

The three-electrode resistivity test platform was used for DC volume resistivity testing, and the DC leakage current was measured by an electrostatic current meter (Keithley 2635b, USA). The average thickness of the sample was 100 μm. The test temperature was 30 °C and 70 °C, respectively, and the electric field strength ranged from 5 kV/mm to 60 kV/mm with the interval of 5 kV/mm. The current values in the stable segment were averaged as the measured value.

The thermally stimulated depolarization currents (TSDC) were used for testing the trap energy level distribution of materials [14]. The average thickness of the sample was 100 μm. The samples were applied electric field strength of 5 kV/mm at a polarization temperature of 70 °C for 30 min, then short-circuited for 5 min after rapidly cooled to −60 °C, and finally heated to 120 °C at the rate of 3 °C/min meanwhile measuring the TSDC.

The HVDC tester was used to test the DC breakdown strength, and the HVAC tester was used to test the AC breakdown strength. The samples with an average thickness of 100 μm were placed between the sphere-sphere electrodes. The electrodes and the sample were immersed in silicone oil for preventing the flashover. The sample was applied voltage at the rate of 1 kV/s until breakdown. The measured values of 15 different breakdown locations were selected in each sample, and statistical analysis was adopted by the two-parameter Weibull distribution [15].

The broadband dielectric spectrometer (BDS), modeled as Novocontrol Concept 80, was used for testing dielectric properties. The average thickness of the sample was 100 μm. The sample was applied 1 Vrms, the frequency from 1 Hz to 10^6 Hz, the temperature of 30 °C and 70 °C respectively.

The pulsed electro-acoustic method (PEA) was used for testing the change of space charge. The average thickness of the sample was 200 μm. The sample was applied a polarized electric field of 30 kV/mm at room temperature, and a polarization time of 40 min.

3. Results and Discussion

3.1. FTIR

The results of infrared spectroscopy were shown in Figure 2. It can be seen that compared with pure PP, new vibration peaks of the benzene ring skeleton of PPg appeared at wavenumbers of 1500 cm^{-1} and 1600 cm^{-1}, as shown in the yellow box. The three C-H bond expansion peaks in benzene rings appeared from 3105 cm^{-1} to 3020 cm^{-1}, as shown in the green box. The vibration peak of the benzene ring skeleton appeared at 700 cm^{-1}, as shown in the blue box. It indicated that the purified PPg material was still able to characterize the characteristic peaks of the styrene groups, which ensured the successful grafting of the styrene groups on PP molecular chains.

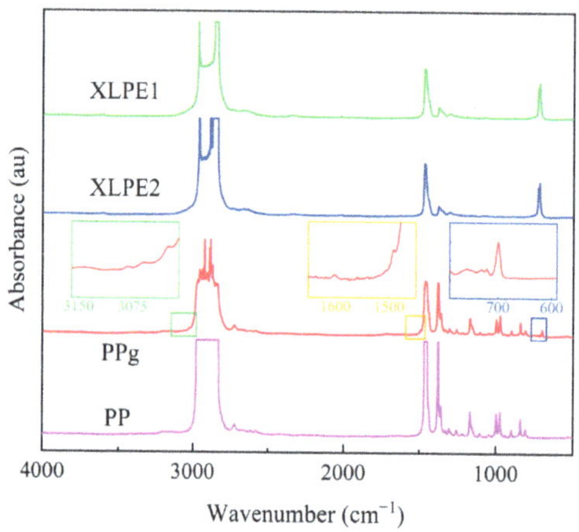

Figure 2. FTIR of the three materials.

3.2. DSC

The melting and crystallization heat flow of the three materials were obtained by the DSC test as shown in Figure 3. The thermal parameters of the materials were shown in Table 2. The crystallinity was calculated by dividing the measured melting enthalpy by 100% melting enthalpy, where 100% melting enthalpy of XLPE1 and XLPE2 was 290 J/g, and that of PPg was 209 J/g [15]. As shown in Figure 3 and Table 2, the melting peak temperature of PPg was significantly higher than that of XLPE1 and XLPE2 by approximately 50 °C, and XLPEs began to absorb heat above 85 °C, which limited the increase of the current-carrying capacity of the cables. The PPg started heat absorption above 135 °C, and the melting peak temperature exceeded 150°C and was approximately 50 °C higher than XLPEs. Because polyethylene (PE) was a semi-crystalline plastic, the crystallinity of XLPE decreases with the increase of crosslinking degree. The crystallinity of PP was generally lower than that of PE, in addition, the steric hindrance effect produced by the grafted benzene ring groups hindered the movement and order of PP molecular chains to a certain extent, and finally decreases the crystallinity.

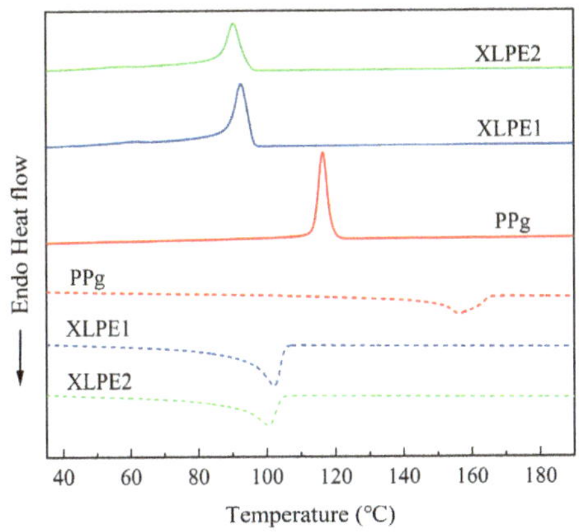

Figure 3. DSC results of three materials.

Table 2. Thermal parameters of three materials.

Thermal Parameters	XLPE1	XLPE2	PPg
The temperature of melting peak (°C)	102.2	100.6	156.3
Temperature of crystallization peak (°C)	92.7	90.6	116.5
Melting enthalpy (J/g)	104.1	102.4	49.9
crystallinity (%)	36.0	35.3	23.9

3.3. Mechanical Tensile Test

The stress-elongation curves obtained by the mechanical tensile test are shown in Figure 4, and the corresponding parameters of mechanical properties are listed in Table 3. As shown in the subplot of Figure 4, the overall tension modulus of PPg was higher than that of XLPEs, due to the weaker intrinsic toughness of the PP matrix above the glass transition temperature. The PPg had no thermosetting cross-linking structure and failed to form a network structure, which made the PP molecular chains more extended. In addition, the grafted styrene groups with a large volume, reduced the intermolecular binding forces, and the molecular chains were more likely to slip, making the break elongation slightly higher than XLPEs. The PPg with low crystallinity, made the reduction of macromolecules subjected to stress during stretching, it also increased the break elongation. In a word, the mechanical tensile properties of PPg could meet the requirements of minimum break elongation of 200% and minimum tensile strength of 12.5 N/mm^2 in cable insulation [16].

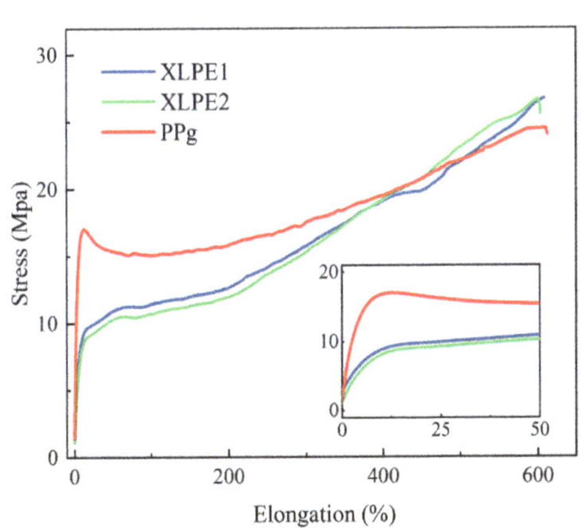

Figure 4. Stress-elongation curves of three materials.

Table 3. Mechanical properties of three materials.

Mechanical Properties	XLPE1	XLPE2	PPG
Tension modulus (MPPa)	120.9 ± 7.4	127.5 ± 7.3	538.3 ± 45.5
Elongation (%)	555.6 ± 19.7	622.1 ± 28.3	652.7 ± 32.3
Tensile strength (N/mm^2)	28.8 ± 7.1	32.5 ± 1.9	31.8 ± 2.1

3.4. TSDC

The thermal stimulation current test obtained the change of current with temperatures, and the relationship of trap density and depth was calculated by the improved TSDC analysis method, the result of which is shown in Figure 5 [14]. The trap depth of PPg was 1.145 eV, XLPE1 of 0.995 eV, and XLPE2 of 1.031 eV. The results showed that PPg had deeper traps and higher density. This is due to the grafted styrene groups introducing aromatic ring structure into the PP molecular chains. The aromatic ring possesses delocalized Pi bond, which is an electron-affinitive structure, therefore it can act as a deep charge trap, meanwhile changing the electronic state distribution. Moreover, the grafted styrene groups affected the crystallization behaviors of PP, the size of the spherulites became smaller and the boundary became blurred. Thus, it increased the interface between the crystalline region and the amorphous region and made the trap quantity increase. A large increase in deep trap density could capture more carriers and reduce the migration of charges, which was conducive to improving the resistivity and breakdown strength [12].

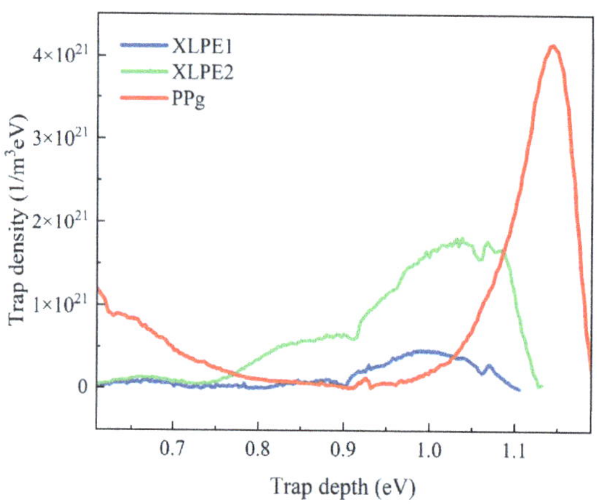

Figure 5. Trap distribution of three materials.

3.5. DC Volume Resistivity

The results of DC volume resistivity were shown in Figure 6, the applied electric field strengths of 5 kV/mm, 15 kV/mm and 25 kV/mm, and the temperatures at 30 °C and 70 °C. The DC volume resistivity of XLPEs, at the high electric field and high temperature, reduced by 2 orders of magnitude from that at low field and high temperature. The volume resistivity of PPg was higher than XLPEs at different field strengths and temperatures, and the reduction rate of that was much lower than XLPEs with the field strength and temperature increasing. It indicated that the volume resistivity of PPg was less dependent on the field strength and temperatures. The higher volume resistivity of PPg could be attributed to the grafted styrene groups introducing deep traps, which captured the carriers and reduce the mobility of carriers [5,11], thereby reducing the leakage current and improving the resistivity, corresponding to the trap level distribution in Figure 5. This indicated that the DC leakage loss of PPg was lower than that of XLPEs at high temperatures, which were beneficial for the safe and stable operations of cables at conditions of high temperatures and high field strength.

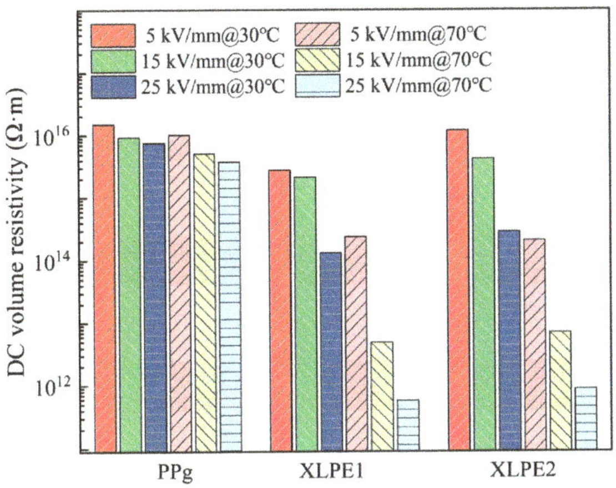

Figure 6. DC volume resistivity of three materials.

3.6. AC and DC Breakdown Strength

The AC and DC breakdown strength can be analyzed by Weibull distribution, as shown in Equation (1):

$$P(E) = 1 - \exp\left(-\left(\frac{E}{E_C}\right)^{\beta}\right) \quad (1)$$

where $P(E)$ is the cumulative breakdown probability, and E is the experimental breakdown strength value. E_C is the fitted characteristic breakdown strength, which corresponds to the electric field strength at the cumulative breakdown probability of 63.2%. The fitted breakdown strength curves of the three materials are shown in Figures 7 and 8, respectively. While the E_C values are shown in Figure 9. The AC breakdown strength of PPg was higher than XLPEs, and it was approximately 34.2% higher than XLPE2 at ambient temperature, and approximately 38.9% higher than XLPE2 at high temperature. The DC breakdown field strength of PPg was also higher than XLPEs, and it was approximately 20.9% and 23.0% higher than that of XLPE1 and XLPE2 at ambient temperature, and approximately 105.2% and 132.3% higher at high temperature respectively. It indicated that PPg still maintained strong DC breakdown resistance at high temperatures, and the breakdown value was close to that of XLPEs at ambient temperature. Meanwhile, the DC breakdown strength of the three materials was higher than the AC breakdown strength of that. Concurrently, it had been discovered that the DC breakdown field strength of the three aforementioned materials exceeded their AC breakdown field strength. This phenomenon was attributed to the homopolar injection of space charges, which provoked gradual movement towards the interior of materials and eventual breakdown. The homopolar injection effectively space charges inhibited further charges injecting. Conversely, during AC breakdown, the alternating electric field resulted in charge injection accumulation at the interface between the medium and electrode, leading to a distortion in local field strength. Furthermore, the repeated trapping and detachment within the materials generated radiation energy, inducing molecular chain breakage and high-energy electron collision ionization, ultimately increasing the current carriers and resulting in AC breakdown [17].

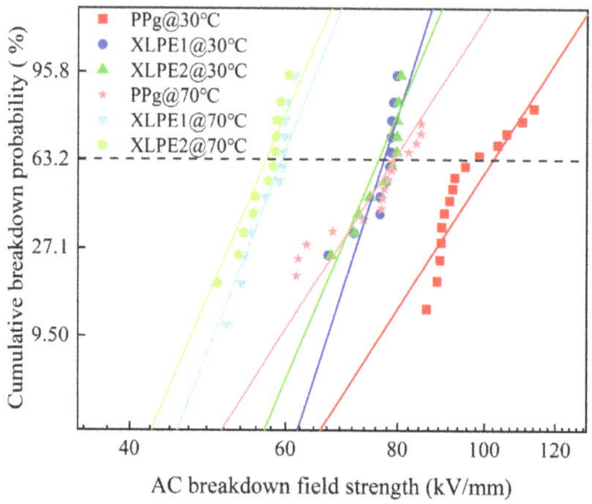

Figure 7. The Weibull distribution of AC breakdown field strength, where the lines are the corresponding fitting curves of Weibull distribution.

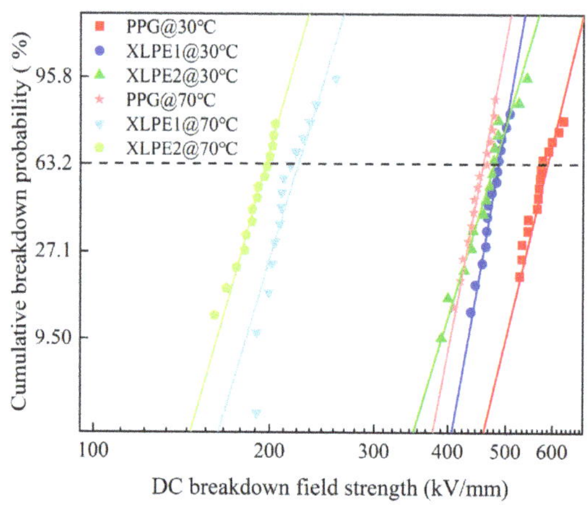

Figure 8. The Weibull distribution of DC breakdown field strength, where the lines are the corresponding fitting curves of Weibull distribution.

Due to the grafted styrene groups in PPg having introduced deep traps (as shown in Figure 5), it captured the charge carriers during the breakdown process, greatly reduced the kinetic energy and mobility thus suppressing the intermolecular energy exchange, thus increasing the DC breakdown field strength [12]. The improvement of AC breakdown field strength, on the one hand, can also be attributed to the introduced deep traps, on the other hand, grafting the rigid styrene groups formed an entangled structure, which enhanced the interaction of molecular chains and thus improved the breakdown properties in high electrical and high thermal fields. It showed that PPg material had excellent breakdown resistance for AC/DC general cable insulation.

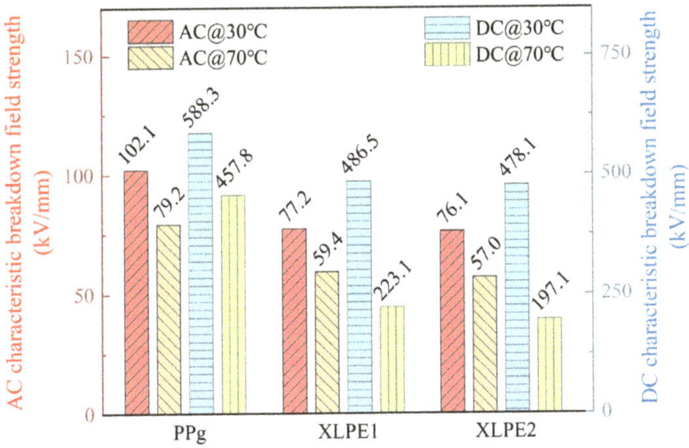

Figure 9. AC and DC characteristic breakdown field strength.

3.7. Dielectric Properties

The present study investigated the frequency-dependent changes in the relative permittivity and dielectric loss tangent (tan δ) of three materials, as shown in Figures 10 and 11. The results indicated that the relative permittivity of the materials exhibited a frequency-dependent behavior and decreased with an increase in temperatures, albeit with a marginal shift in the curve. The rise of temperatures did not alter the general shape of the tan δ curves, but only caused a rightward shift along the frequency axis. Notably, the tan δ value of PPg displayed an initial increase followed by a decrease with an increase in frequency, with a crossing point observed around the power frequency of 50 Hz. In the high-temperature and low-frequency range (below 50 Hz), the tan δ value of PPg was lower than its corresponding ambient temperature value, while it was higher than the ambient temperature tan δ value in the medium-high frequency range (above 50 Hz). Similarly, the tan δ of XLPEs increased initially and then decreased with frequency, with the peak of the curve observed at a frequency level of 10^4 Hz.

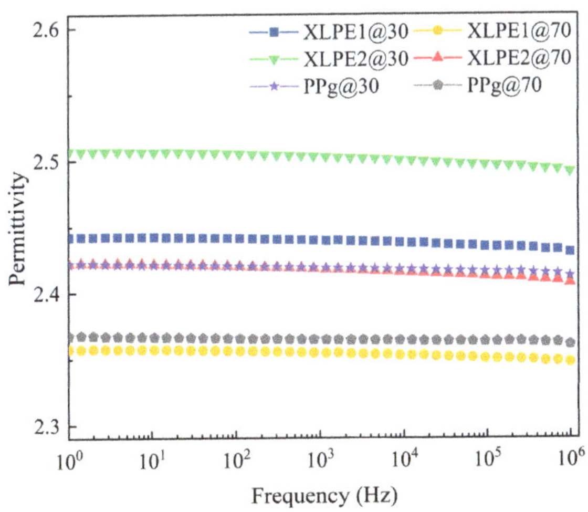

Figure 10. Permittivity varies with frequency.

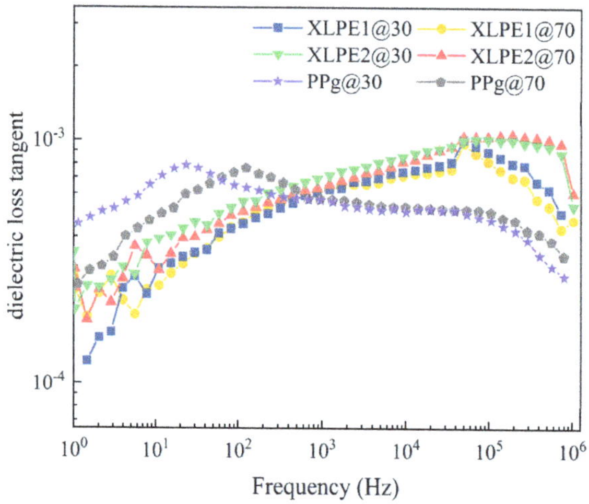

Figure 11. Dielectric loss tangent varying with frequency.

The observed dielectric loss in the materials was attributed to the conduction and polarization losses. The presence of styrene groups grafted in PPg enhanced the orientation and interfacial polarization losses in the low-frequency range. This results in a slightly higher tan δ value for PPg compared to XLPEs in this frequency range, with the difference increasing with frequency. In the medium-high frequency range, the interfacial polarization strength weakened, and the dipole orientation polarization gradually dominates the dielectric response behavior. The entanglement effect of the grafting chains formed by the styrene groups suppressed the movement of the molecular chains, thereby weakening the dipole orientation polarization and causing a significant decrease in the tan δ value of PPg in the high-frequency range [18]. At high temperatures, the increased disorder and intense movement of molecular chains made it challenging to establish polarization, resulting in a decrease in relative permittivity and dielectric loss. As shown in Figure 11, the relative permittivity of PPg at 30 °C and 70 °C was approximately 2.42 and 2.36, respectively, with tan δ values at 50 Hz of 7.02×10^{-4} and 6.79×10^{-4}. The tan δ value of XLPEs exhibited less temperature dependence, and the collision ionization generated with increasing frequency increased the carriers, resulting in an increase in the tan δ value of XLPEs.

3.8. Space Charge

It illustrated the distribution of space charges density and electrical field distribution in the thickness direction of the three materials at ambient temperature in Figure 12. Among them the electric field distribution is calculated by Poisson Equation (2) [5]:

$$E(x) = \frac{1}{\varepsilon_r \varepsilon_0} \int_0^d \rho(x) \mathrm{d}x \tag{2}$$

where ε_0 is the vacuum permittivity, and the ε_r is the relative permittivity of the sample. d is the thickness of the sample. $\rho(x)$ is the distribution of space charge obtained by PEA, while $E(x)$ is the electric field distribution. It was not observed the significant accumulation of space charges or the distortion of the electrical field in PPg material. However, XLPE1 showed a small accumulation of the same polarity charges, which intensified the field distortion near the electrode. On the other hand, XLPE2 showed a small accumulation of opposite polarity charges, and the degree of field distortion was slightly lower than that in XLPE1 due to the charge's recombination. The accumulation of opposite polarity charges in XLPEs was attributed to by-products generated during the cross-linking reaction, while the accumulation of the same polarity charges was mainly caused by electrode injection [19]. The styrene groups grafted in PPg exhibited an excellent ability to suppress the accumulation of space charges and field distortion. It was due to the introduction of a large number of deep traps after the grafting styrene groups, which captured carriers to reduce carriers mobility. Additionally, the captured charges can form a local potential barrier which can suppress the following injection of electrode charges. The ability to suppress the accumulation of space charges could prevent local degradation problems caused by charges accumulation in DC cable insulation, thereby improving the safe operation of DC cables.

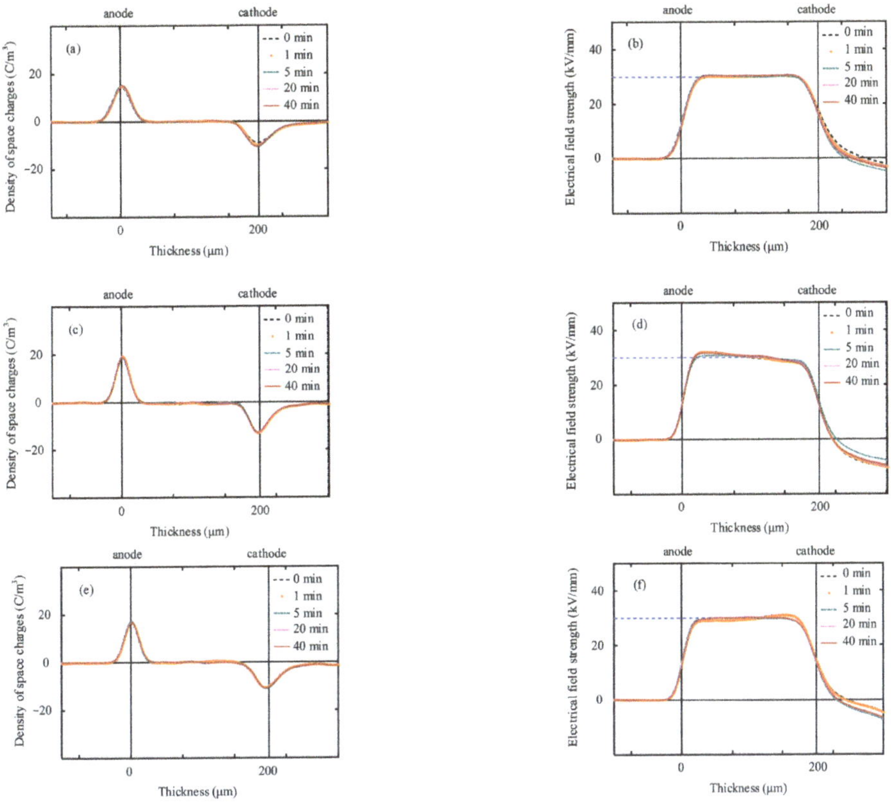

Figure 12. Characteristics of space charges. (**a–f**) are the density of space charges and electrical field strength in PPg, XLPE1, and XLPE2, respectively.

4. Conclusions

It was studied that grafted polypropylene insulation materials for AC/DC distribution power cables, and their comprehensive properties were evaluated. The results demonstrated that:

(1) The PPg has possessed superior AC/DC electrical properties compared to XLPE materials, including higher DC volume resistivity at high temperatures, higher breakdown field strength, and a better suppression of space charges. The volume resistivity of PPg at high temperatures was found to be 1 to 2 orders of magnitude higher than that of XLPEs. Additionally, the characteristic breakdown field strength of PPg was higher than that of XLPEs, with the DC characteristic breakdown field strength of PPg at high temperatures being over 100% higher than that of XLPEs. The permittivity and dielectric loss of PPg at high temperatures were slightly lower than those at ambient temperature, and showed different dielectric properties from XLPEs. Furthermore, the degree of space charge accumulation and electric field distortion in PPg were weaker than that in XLPEs.

These findings indicated that PPg exhibited excellent material performances in AC/DC distribution power cables at high temperatures and high field conditions.

(2) The PPg has possessed weaker mechanical properties compared to XLPE materials, and with a higher tension modulus. Further research is needed to focus on optimizing the matching of mechanical and electrical properties. However, the mechanical properties of PPg were found to meet the relevant standard requirements for the operation of power cables.

Author Contributions: Conceptualization, S.H. (Shangshi Huang) and J.H. (Jinliang He); methodology, S.H. (Shangshi Huang); software, J.H. (Jun Hu) and Q.L.; validation, Y.Z., J.Y. and C.Y.; formal analysis, H.Y.; investigation, S.H. (Shangshi Huang); resources, J.H. (Jinliang He); data curation, S.H. (Shangshi Huang); writing—original draft preparation, S.H. (Shangshi Huang); writing—review and editing, S.H. (Shixun Hu); visualization, J.H. (Jinliang He); supervision, Y.Z.; project administration, C.Y.; funding acquisition, J.Y. All authors have read and agreed to the published version of the manuscript.

Funding: This research was funded by the Science and Technology Project of State Grid Corporation of China, grant number 5500-202228113A-1-1-ZN.

Data Availability Statement: Not applicable.

Acknowledgments: The authors extend their appreciation to Qing Shao, Mingti Wang, Yaru Zhang, and Juan Li in SINOPEC Beijing Research Institute of Chemical Industry, and Yunfeng Shi and Baojun Zhu in Baosheng High Voltage Cable Co., Ltd. for providing support.

Conflicts of Interest: The authors declare no conflict of interest.

References

1. Zhou, Y.; Peng, S.; Hu, J.; He, J. Polymeric Insulation Materials for HVDC Cables: Development, Challenges and Future Perspective. *IEEE Trans. Dielectr. Electr. Insul.* **2017**, *24*, 1308–1318. [CrossRef]
2. Huang, X.; Zhang, J.; Jiang, P.; Tanaka, T. Material progress toward recyclable insulation of power cables part 2: Polypropylene based thermoplastic materials. *IEEE Electr. Insul. Mag.* **2020**, *36*, 8–18. [CrossRef]
3. Zhang, L.; Wang, Z.; Tian, J.; Meng, S.; Zhou, Y. Thermal Aging Properties of 500 kV AC and DC XLPE Cable Insulation Materials *Polymers* **2022**, *14*, 5400. [CrossRef] [PubMed]
4. Kim, D.; Lee, S.H.; Kwon, T.H.; Kwon, I.; Han, D.H.; Park, H.; Han, S.W.; Lee, D.H.; Yu, S. Study on high-temperature and high voltage insulation characteristics of polypropylene blend with highly packed elastomeric domains for power cable applications *Polym. Test.* **2023**, *120*, 107942. [CrossRef]
5. Hu, S.; Zhou, Y.; Yuan, C.; Wang, W.; Hu, J.; Li, Q.; He, J. Surface-modification effect of MgO nanoparticles on the electrical properties of polypropylene nanocomposite. *High Volt.* **2020**, *5*, 249–255. [CrossRef]
6. Zhou, Y.; He, J.; Hu, J.; Huang, X.; Jiang, P. Evaluation of polypropylene/polyolefin elastomer blends for potential recyclable HVDC cable insulation applications. *IEEE Trans. Dielectr. Electr. Insul.* **2015**, *22*, 673–681. [CrossRef]
7. Bai, L.; Fan, D.; Li, T.; Li, B.; Su, M.; Fan, S.; Zhang, L. Influence of surface discharge on the deterioration characteristics of ethylene-propylene rubber cable insulation under alternating current high voltage. *IET Sci. Meas. Technol.* **2022**, *16*, 293–304 [CrossRef]
8. Qin, S.; Li, Q.; He, M.; Shao, H.; Yu, J.; Guo, J.; Zhang, K.; Yan, W. Study on Thermal Stability and Flame Retardancy of Polymer/Layered Silicate Nanocomposites Based on POE and POE-g-MAH. *Polym. Eng. Sci.* **2014**, *54*, 2911–2917. [CrossRef]
9. Dang, B.; Li, Q.; Zhou, Y.; Hu, J.; He, J. Suppression of elevated temperature space charge accumulation in polypropylene/elastomer blends by deep traps induced by surface-modified ZnO nanoparticles. *Compos. Sci. Technol.* **2017**, *153*, 103–110 [CrossRef]
10. Krentz, T.; Khani, M.M.; Bell, M.; Benicewicz, B.C.; Nelson, J.K.; Zhao, S.; Hillborg, H.; Schadler, L.S. Morphologically dependent alternating-current and direct-current breakdown strength in silica-polypropylene nanocomposites. *J. Appl. Polym. Sci.* **2017**, *134* 44347. [CrossRef]
11. Hu, S.; Zhang, W.; Wang, W.; Li, J.; Shao, Q.; Zhang, Y.; Zhang, Q.; Huang, S.; Hu, J.; Li, Q.; et al. Comprehensive Comparisons of Grafting-Modified Different Polypropylene as HVDC Cable Insulation Material. *IEEE Trans. Dielectr. Electr. Insul.* **2022**, *29* 1865–1872. [CrossRef]
12. Yuan, H.; Zhou, Y.; Zhu, Y.; Hu, S.; Yuan, C.; Song, W.; Shao, Q.; Zhang, Q.; Hu, J.; Li, Q.; et al. Origins and effects of deep traps in functional group grafted polymeric dielectric materials. *J. Phys. D* **2020**, *53*, 475301. [CrossRef]
13. Chmela, S.; Fiedlerova, A.; Janigova, I.; Novak, I.; Borsig, E. Grafting of iPP Powder with Methacrylate Monemers in Water Medium. *J. Appl. Polym. Sci.* **2011**, *119*, 2750–2758. [CrossRef]

14. Tian, F.; Bu, W.; Shi, L.; Yang, C.; Wang, Y.; Lei, Q. Theory of modified thermally stimulated current and direct determination of trap level distribution. *J. Electrostat.* **2011**, *69*, 7–10. [CrossRef]
15. Zhang, C.; Wang, T.; Li, C.; Zhao, H.; Wang, X. emperature-Dependent Space Charge and Breakdown Strength of Aromatic Amine Antioxidant Grafted Crosslinked Polyethylene. *IEEE Trans. Dielectr. Electr. Insul.* **2023**, *30*, 56–64. [CrossRef]
16. *IEC 60502-2*; Power Cables with Extruded Insulation and Their Accessories for Rated Voltages from 1 kV (Um = 1, 2 kV) Up to 30 kV (Um = 36 kV)-Part 2: Cables for Rated Voltages of 6 kV (Um = 7.2 kV) and 30 kV (Um = 36 kV). International Electrotechnical Commission: London, UK, 2021.
17. Du, B.; Zhang, J.; Xiao, M.; Liu, H.; Ran, Z. Dielectric Breakdown Performance of Deashed Polypropylene for HVDC Film Capacitors. *IEEE Trans. Dielectr. Electr. Insul.* **2023**, *29*, 2209–2217. [CrossRef]
18. Lee, S.H.; Kim, D.; Kwon, I.; Kwon, T.H.; Park, H.; Han, D.H.; Han, S.W.; Lee, J.H.; Lee, D.H.; Yu, S. Polystyrene: A Self-Dispersing, Ultralow Loading Additive for Improving the Breakdown Strength of Polypropylene for High Voltage Power Cable Applications. *ACS Appl. Polym. Mater.* **2023**, *5*, 165–171. [CrossRef]
19. Hussin, N.; Chen, G. Analysis of space charge formation in LDPE in the presence of crosslinking byproducts. *IEEE Trans. Dielectr. Electr. Insul.* **2012**, *19*, 126–133. [CrossRef]

energies

MDPI

Article

Impact of Air Gap Defects on the Electrical and Mechanical Properties of a 320 kV Direct Current Gas Insulated Transmission Line Spacer

Yuan Deng [1,2], Xianhao Fan [3,*], Hanhua Luo [3], Yao Wang [4], Keyan Wu [4], Fangwei Liang [3,*] and Chuanyang Li [3,*]

1 Department of Electrical Engineering, Xi'an Jiaotong University, Xi'an 710049, China
2 PingGao Group Co., Ltd., Pingdingshan 467001, China
3 Department of Electrical Engineering, Tsinghua University, Beijing 100084, China
4 Tai'an Taishan High Voltage Switchgear Co., Ltd., Tai'an 271000, China
* Correspondence: xianhao_fan@163.com (X.F.); liangfangwei@tsinghua.edu.cn (F.L.); chuanyang_li@mail.tsinghua.edu.cn (C.L.)

Abstract: Air gap defects inside a spacer reduce its insulation performance, resulting in stress concentration, partial discharge, and even flashover. If such gap defects are located at the interface between the insulation and conductor, a decrease in mechanical stress may occur. In this work, a finite element method-based simulation model is developed to analyze the influence of gap defects on the electrical and mechanical properties of a ±320 kV direct current gas insulated line (DC GIL) spacer. Present findings reveal that a radially distributed air gap produces a more significant effect on the electric field distribution, and an electric field strength 1.7 times greater than that of the maximum surface value is observed at the air gap. The axial distribution dominates the distortion of the surface stress by generating a stress concentration region in which the maximum stress of the air gap is twice the pressure in the surrounding area.

Keywords: gas insulated line; insulation spacer; finite element method; air gap defect; simulation analysis

1. Introduction

With its advantages of low loss, large transmission capacity, and easy grid interconnection, high voltage direct current (HVDC) transmission is steadily influencing the future direction of modern power systems [1–3]. A GIL spacer offers unique advantages and is thus widely used to pass through partial areas with large vertical drops and poor meteorological conditions [4–6]. Among the components of a GIL, the insulation spacer is indispensable for tasks such as isolating the air chamber, electrical insulation, and mechanical support [7,8].

The GIL is filled with SF_6 gas [9] with a pressure of 0.45 MPa during operation. Obvious internal stress concentration may occur under the external stress load or around micro defects, which makes this region the weakest point of the spacer. However, bubbles and other defects can be inserted into the spacer as a result of imperfect manufacturing techniques or unstable equipment during production [10,11]. A micro defect can distort the electric field of the spacer [12,13], which leads to partial discharge or surface flashover. Compared to the AC spacer, the DC spacer offers relatively insufficient mechanical properties due to its flatter geometry. Accumulation of spatial charge under DC conditions further significantly reduces the flashover voltage of the DC spacer [14]. However, research thus far has focused mainly on charge accumulation and charge suppression and less on the influence of interior defects on the electrical or mechanical properties of DC spacers [1,14]. Therefore, the potential factors leading to a decline in mechanical and electrical performance under DC conditions need attention.

The specific stress on a spacer can be obtained in a mechanical water pressure test by arranging strain sensors on its surface to measure the stress at each position [15].

Measuring electric field intensity is more complex and mainly involves an electric field probe [16,17]. However, the probe will distort the electric field distribution, causing measurement errors. Other methods, such as dust mapping and electroluminescence [18], are also used, but it is difficult to obtain quantitative results by image reflection. It is also difficult to accurately analyze the mechanical stress and electric field intensity distribution by merely depending on experimentation. This issue can be overcome using the finite element method (FEM) [19,20]. According to the geometric structure, stress environment, and boundary conditions of the physical object, a simulation model can be developed to promote accurate and real-time analysis of the electrical and mechanical properties of an insulation spacer. One study [21] used the FEM method to simulate an insulation spacer under a hydrostatic test to determine the strain and stress distributions along the spacer surface. Another study examined the effect of adhering spherical conducting particles (with different sizes and locations) on the electric field distribution [22]. The group in [23] analyzed the electric field properties under two types of interface gaps via a 2D model. The literature [24] investigates the partial discharges raised by floating particles and nitrogen bubbles with different shapes and radii. All of these studies show the feasibility and applicability of the FEM. However, a more comprehensive spatial model including both electrical and mechanical properties should be constructed to investigate the field distortion.

Given the need for such a model, an FEM-based simulation model is constructed to investigate the electric field intensity and mechanical stress under various radially and axially distributed air gap defects. The variation in electric field intensity and stress versus radial or axial distance on the spacer surface is analyzed and discussed, and the corresponding functional relationship is established. Present findings reveal that a radial air gap influences the electric field distribution more significantly than an axially distributed air gap that dominates the surface stress deformation in different ways. The conclusions benefit a quantitative understanding of the impact of DC spacers on electrical and mechanical properties due to various air gap defects.

2. Simulation Model

2.1. Geometry

In this work, a defect spacer is researched to determine how an air gap impacts the GIL insulation performance. The FEM is used to achieve this purpose. Referring to previous work [25], the geometry is shown in Figure 1 and contains three types of material. The detailed material and geometric parameters are listed in Table 1.

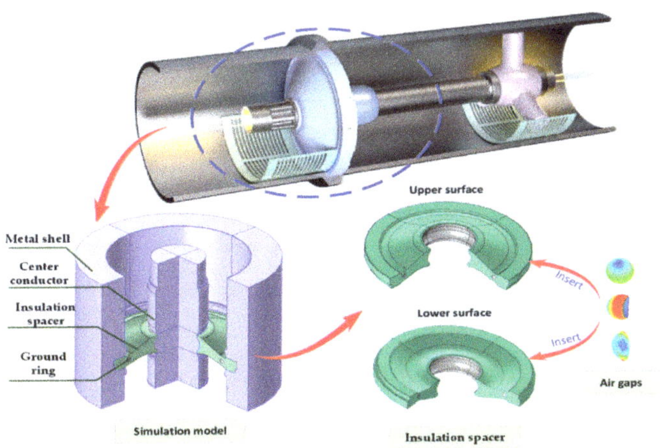

Figure 1. GIL simulation model.

Table 1. Geometric and material parameters of the simulation model.

No.	Component	Geometry Parameter	Material	Material Parameter
1	Metal shell	Length: 52 cm Radius: 33 cm	Aluminum alloy	Relative permittivity: 1×10^7 Young modulus: 71 GPa Poisson's ratio: 0.33 Density: 2700 kg/m^3
2	Center conductor	Length: 52 cm Radius: 9.8 cm		
3	Ground ring	Radius: 1 cm		
4	Insulation spacer	Max thickness: 5 cm Radius: 27 cm	Epoxy resin	Relative permittivity: 4.95 Young modulus: 13 Gpa Poisson's ratio: 0.36 Density: 2300 kg/m^3
5	Air gap	Radius: 0.15 cm	Air	Relative permittivity: 1

2.2. Equation Derivation and Boundary Conditions

The insulation spacer in Figure 1 is operated at DC 320 kV. A DC voltage is preset at the center conductor (labeled 2 in Table 1), then the electric field intensity is generated and distributed along the upper and lower surfaces of the insulation spacer (labeled 4).

Given that the surface electric field intensity (E) is required to be less than 12 kV/mm under normal operation, making the monitor of this index is rather fundamental. In the simulation model, E can be also defined in terms of the negative potential φ:

$$E = -\nabla\varphi \tag{1}$$

When the metal shell and ground ring are grounded (labeled 1 and 3 in Table 1), the potential in the region between them satisfies

$$\nabla \cdot (\nabla\varphi) = 0 \tag{2}$$

The differential equations corresponding to the adjacent interfaces can be defined to describe transferred properties [26]. The subscript indicates each interface, n, is a normal vector, and ε is the relative permittivity. When the potential distribution in the computation regions is determined, the corresponding E can be naturally obtained via (1).

$$\begin{array}{c} \varphi_1 = \varphi_2 \\ \varepsilon_1 \frac{\partial\varphi_1}{\partial n} = \varepsilon_2 \frac{\partial\varphi_2}{\partial n} \end{array} \tag{3}$$

In addition to the necessary electrical performance, excellent mechanical performance is also required of a GIL insulation spacer because it always operates under a pressure of 0.45 MPa in a SF_6 atmosphere. To check the damage to the insulation spacer caused by the air gap, a 2.4-MPa pressure load is preset on the lower surface according to a hydrostatic test [20]. Once the external load is applied, all points inside the insulation spacer are in a state of stress balance.

During the simulation, the intersection between each spacer and the metal shell and between each spacer and the central conductor is set as a fixed constraint, and the upper surface and ground ring are treated as free boundaries. The epoxies filling the insulation spacer can be treated as isotropic materials, which means the Young moduli in all directions (G_{xy}, G_{xz}, G_{yz}) are the same:

$$G_{xy} = G_{xz} = G_{yz} \tag{4}$$

Finally, the stress boundary condition of the computation region Γ is obtained from the sum of boundary conditions of the external stress Γ_σ and internal deformation Γ_u:

$$\Gamma = \Gamma_\sigma + \Gamma_u \tag{5}$$

2.3. Performance Analysis with a Defect-Free Model

A total of 3,644,364 degrees of freedom were generated during the simulation. Figure 2 presents results of mesh generation of the simulation results. Figure 3 presents the distributions of the electric field intensity (modulus) and principal stress of the insulation spacer and its cross section without any defect. Figure 3b,d provide the cross profiles of Figure 3a,c, which are formed by rotating 360 degrees with the Z axis as the symmetry axis in Figure 3b,d.

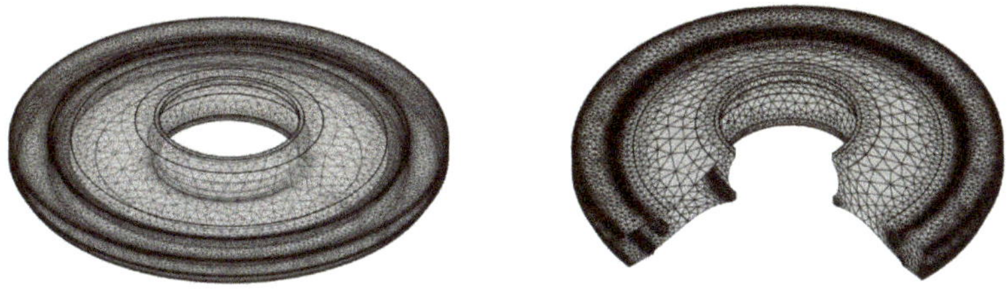

Figure 2. Results of mesh generation of the simulation results.

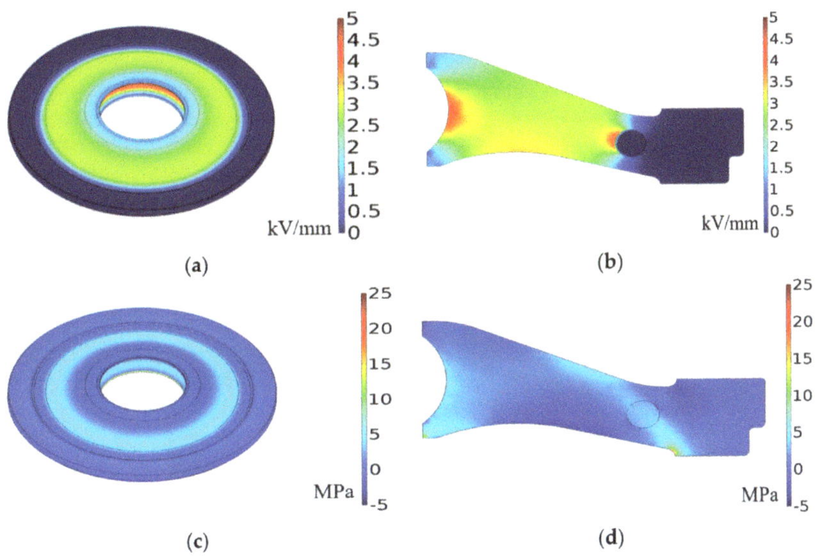

Figure 3. Simulated distributions of (**a**,**b**) electric field intensity, and (**c**,**b**) first principal stress of the defect−free simulation model.

Table 2 shows the maximum electric field intensities and principal stresses (E_{max} and P_{max}) corresponding to the upper and lower surfaces of the insulation spacer. The electrical and mechanical performances of the defect-free model are sufficient for normal operation. To show how mechanical stress affects the spacers, Figure 4a describes the distribution of mechanical forces around the surface of the spacer, and Figure 4b presents the resulting shape of the spacer when conditioning the deformation caused by mechanical forces.

Table 2. Electrical and mechanical performances of an insulation spacer without any defect.

Items	E_{max} (kV/mm)	P_{max} (MPa)
Upper surface	2.929	14.365
Lower surface	3.427	25.632

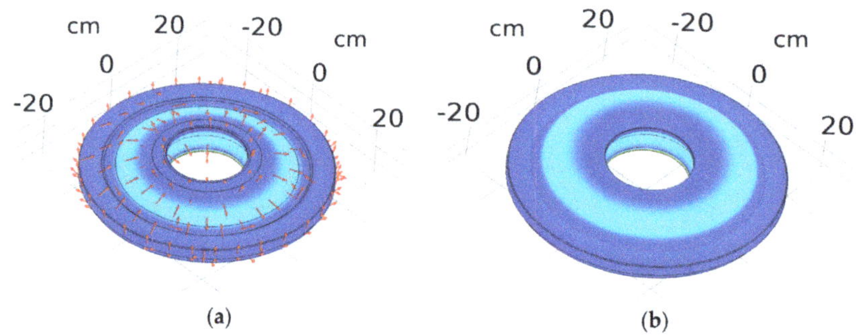

Figure 4. Simulated distributions of (**a**) mechanical forces and (**b**) deformation of the defect-free simulation model.

3. Impact on Model Performance of a Radially Distributed Air Gap Defect

3.1. Arrangement of the Air Gap Defect along the Radial Direction of the Insulation Spacer

In this section, the air gap is arranged inside the insulation spacer and has a radial distribution, as shown in Figure 5. The geometry of the air gap is simulated as a sphere with a radius of 3 mm. The air gap caused by the poor wetting operation is mostly spherical in the manufacturing process. Also, there is no uniform standard for the air gap shape caused by other faults, thus the simple sphere is utilized. In addition, the size of the air gap would not sharply alter the distribution of the electric field compared to that with diverse radii [24]. The size of the air gap generated in the manufacturing process is generally not too large, otherwise, it cannot pass the production inspection, while defects of too small a size cannot be observed. Therefore, a sphere with a radius of 3 mm is selected. Finally, ten equally spaced positions are distributed to install an air gap with similar distances between these positions and the upper and lower surfaces, The involved material parameter is permittivity, which is shown in Table 1. The distance between the center of the air gap and the conductor is defined as the radial distance *X*-dis and is in the unit mm.

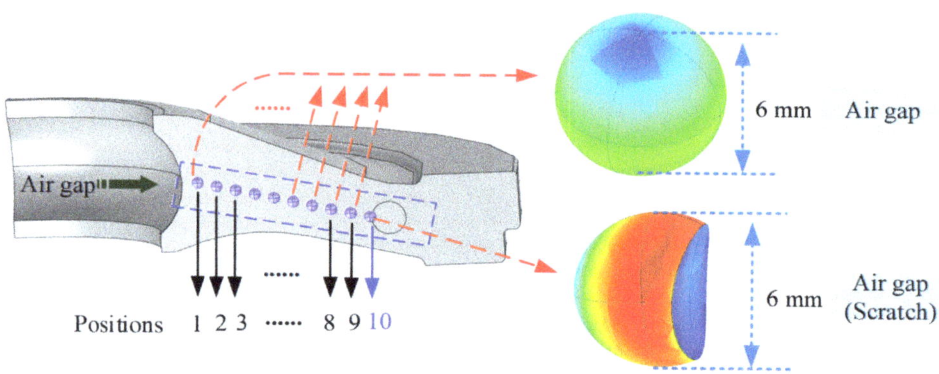

Figure 5. Radial air gap arrangement inside the insulation spacer.

The spacing between positions is 1 mm in Figure 5. The corresponding axial coordinate varies as a function of *X-dis* to ensure similar distances of the upper and lower surfaces of the insulation spacer. There is an intersection between the ground ring and the air gap marked 10. The geometry of this air gap can be altered at the interface, and the air gap can also be treated as a suspended contact defect (like a scratch on the surface). Ten simulations were sequentially carried out to analyze the distributions of electric field intensity and principal stress of the insulation spacer.

3.2. Impact on Electric Field Distribution

The electric field distribution and its maximum (E_{max}) on the air gap surface are tabulated in Table 3. The minimum electric field is 4.144 kV/mm, and the maximum appears at the start and end points of this distribution line. This range reflects the intersection of the spacer with the conductor and ground ring.

Table 3. Electric field distribution and its maximum value on the air gap surface.

X-dis	11 mm	12 mm	13 mm	14 mm	15 mm
E_{max}	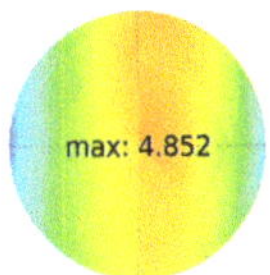	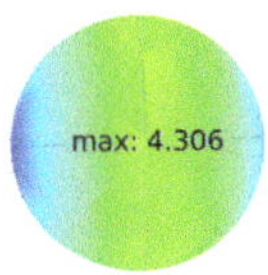			
X-dis	16 mm	17 mm	18 mm	19 mm	20 mm
E_{max}					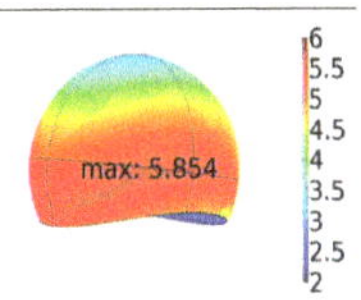

For each air gap, the electric field intensity on both ends is relatively low compared with that of the remaining region, and its maximum is always at the center. The air gap sharply increases around this position and causes a discharge breakdown once the electric field intensity is more than 3 kV/mm (the breakdown voltage of air) [23]. In this case, the accumulated discharge state gradually ages the epoxy materials and generates more charge carriers, which in turn promotes the discharge intensity. Consequently, once the air gap appears inside the insulation spacer, the maximum electric field strength on the air gap surface can lead to breakdown and trigger partial discharge regardless of its radial distance from the conductor. The discharge severity is positively related to the distance between the air gap and the conductor or grounding terminal, as illustrated by the points at 11 mm and 20 mm. The distorted air gap can be treated as a surface scratch defect when it intersects the ground ring. The distorted electric field intensity at 20 mm even exceeds 5.8 km/mm and is 1.2 times larger than that at 11 mm. This result reveals that the suspension discharge maintains a stronger intensity than that of the air gap discharge in the insulation spacer.

Figure 6 shows the distribution level along the surface and cross profile of the spacer when the air gap is treated as a surface scratch. The resulting maximum electric field intensity appears on the bottom surface and develops into the weakest point to form the defect in the insulation spacer. Figure 7 compares the E_{max} in the air gap with that on the upper and lower surfaces of the spacer and with that of a defect-free spacer. If the distance between the air gap and the upper and lower surfaces is sufficient, E_{max} for each surface is consistent with the defect-free state and is not affected by the air gap. The plot of E_{max} versus radial distance (*X-dis*) is fitted in Figure 7b to quantitatively analyze the distribution of E_{max}. Table 4 lists the resulting amplitude ratios of E_{max} corresponding to the air gap (E_{max_air}) and its upper and lower surfaces (E_{max_upperI} and E_{max_lower}) according to:

$$
\begin{aligned}
k_{e\_upper} &= E_{\max\_air} / E_{\max\_upper} \\
k_{e\_lower} &= E_{\max\_air} / E_{\max\_lower}
\end{aligned}
\tag{6}
$$

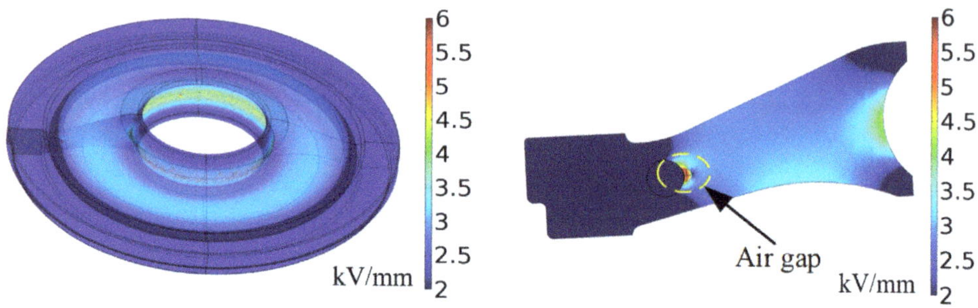

Figure 6. Electric field intensity distribution of an air gap at 20 mm.

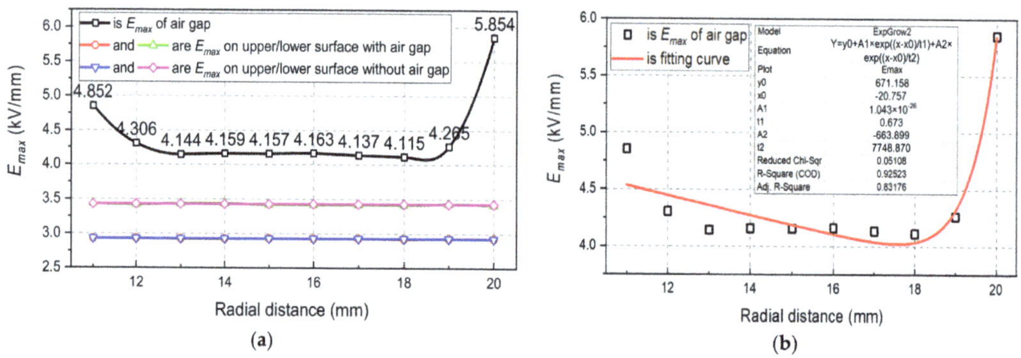

Figure 7. (**a**) Electric field intensity at different positions. (**b**) Fitting analysis of electric field intensity and radial distance.

Table 4. Amplitude ratios of electric field intensity of the air gap for the upper and lower surfaces.

X-dis	11 mm	12 mm	13 mm	14 mm	15 mm
k_{e_upper} k_{e_lower}	1.657 1.416	1.470 1.257	1.415 1.209	1.420 1.214	1.419 1.213
X-dis	16 mm	17 mm	18 mm	19 mm	20 mm
k_{e_upper} k_{e_lower}	1.421 1.215	1.412 1.207	1.405 1.201	1.456 1.244	1.999 1.708

3.3. Impact on Mechanical Stress Distribution

The insulation spacer manufactured with epoxy material has a tensile stress failure threshold of 70 MPa, and the tensile stress at the bond between the spacer and the conductor is 25 MPa. Consequently, a lower mechanical stress amplitude is required to ensure safe operation [20]. Only the lower surface sustains the atmospheric pressure, so once the single-side compression is considered, mechanical stress concentration should be avoided as soon as possible. The distribution and maximum value (P_{max}) of the major principal stress on the air gap surface are shown in Table 5.

Table 5. Mechanical stress distribution and maximum value on the air gap surface.

X-dis	11 mm	12 mm	13 mm	14 mm	15 mm
P_{max}	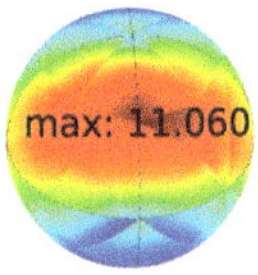 max: 11.060	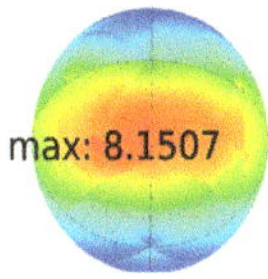 max: 8.1507	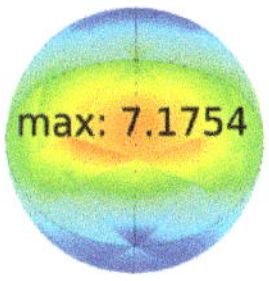 max: 7.1754	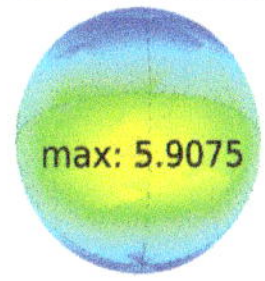 max: 5.9075	 max: 4.4720
X-dis	16 mm	17 mm	18 mm	19 mm	20 mm
P_{max}	 max: 2.5889	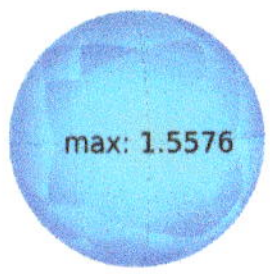 max: 1.5576	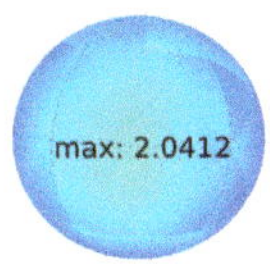 max: 2.0412	 max: 4.2523	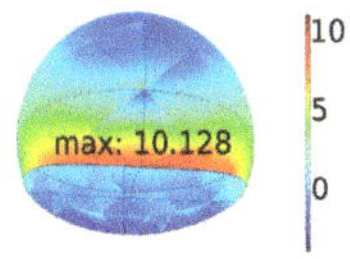 max: 10.128

The observed variation is that P_{max} on the surface of the air gap decreases with increasing radial distance and gradually rises when it is close to the ground ring, similar to the electric field distribution. The distortion can still be observed at the air gap at a radial distance of 20 mm, corresponding to the scratch. The distribution of this scratch is presented in Figure 8, where P_{max} appears at the interface between the lower surface and the metal shell. The same phenomenon is also observed in the other nine cases. The lower surface is believed to be the boundary for sustaining the atmospheric pressure of 2.4 MPa. The generation of the air gap does not change the distribution of the surrounding area as sharply as the electric field intensity. Figure 9a further compares P_{max} on the air gap with that on the upper and lower surfaces of the spacer, with or without the preset defect. The curve of P_{max} on the air gap versus radial distance is then fitted in Figure 9b to highlight the radial distance dependence of the mechanical stress distribution.

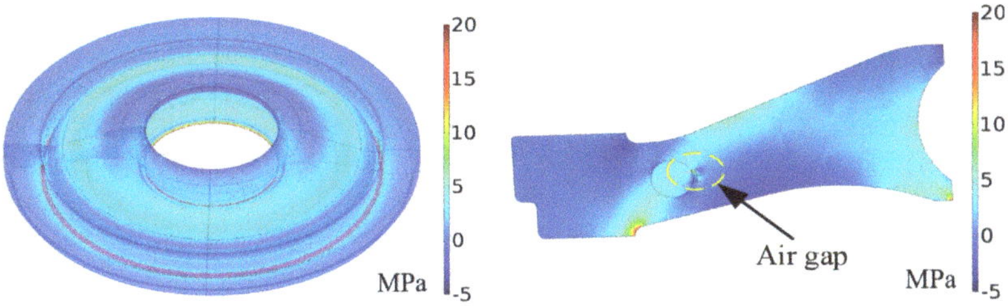

Figure 8. Mechanical stress distribution of an air gap at 20 mm.

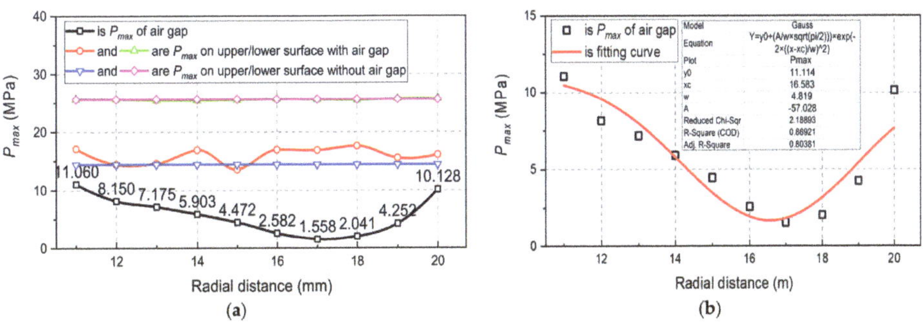

Figure 9. (**a**) Mechanical stress at different positions. (**b**) Fitting analysis of mechanical stress and radial distance.

The P_{max} on the upper and lower surfaces of a defect or a defect-free model is approximately uniform and much higher than that of the air gap. With the air gap, P_{max} on the upper surface varies because it is assigned as the free boundary. The impact of P_{max} on the maximum stress on the upper or lower surface is quantified as the following:

$$\begin{cases} k_{p\_upper} = P_{\max\_air} / P_{\max\_upper} \\ k_{p\_lower} = P_{\max\_air} / P_{\max\_lower} \end{cases} \tag{7}$$

The results are listed in Table 6. As with Figure 8, the mechanical stress is concentrated in neither the air gap nor the inside region but is mainly distributed along the upper and lower surfaces, especially the intersection point.

Table 6. Amplitude ratio of mechanical stress of an air gap compared with the ratios for the upper and lower surfaces.

X-dis	11 mm	12 mm	13 mm	14 mm	15 mm
k_{e_upper}	0.770	0.567	0.499	0.411	0.311
k_{e_lower}	0.431	0.318	0.280	0.230	0.174
X-dis	16 mm	17 mm	18 mm	19 mm	20 mm
k_{e_upper}	0.180	0.108	0.142	0.296	0.705
k_{e_lower}	0.101	0.061	0.080	0.166	0.395

4. Impact on Model Performance under an Axially Distributed Air Gap Defect

4.1. Arrangement of an Air Gap Defect along the Axial Direction of the Insulation Spacer

In the last section, a radially distributed air gap defect was studied to discuss the distortion of the electric field intensity and mechanical stress around the defect. The global maximum of the electrical indicator appeared near the interface between the central conductor and the insulation spacer. Thus, assigning the position at 11 mm as the radial coordinate, we next investigate distortions caused by an axially distributed air gap by changing the axial coordinate from 15 to 22 mm.

In contrast with the radial distribution, the first and last points of the axially distributed air gap form a surface scratch, and the corresponding geometry is determined by the interfaces of the upper and lower surfaces of the spacer. The generated surface scratch in Figure 10 is more representative than that of the suspended contact in Figure 5. Eight air gaps are arranged inside the insulation spacer, as shown in Figure 10.

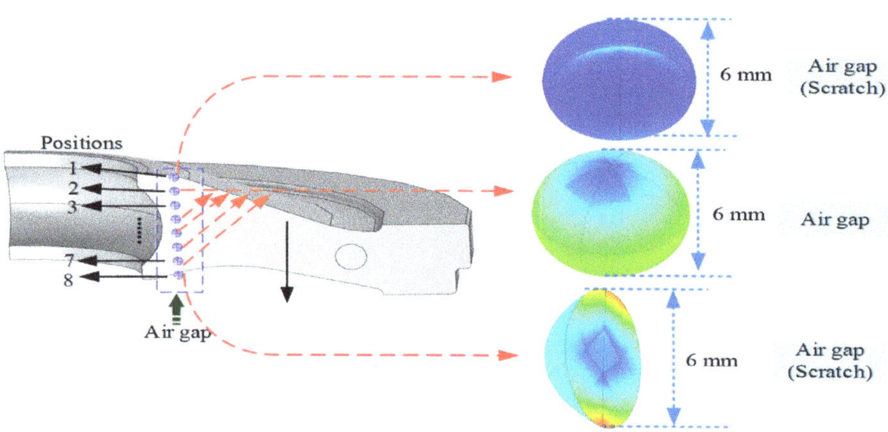

Figure 10. Axial air gap arrangement inside the insulation spacer.

4.2. Impact on Electric Field Distribution

The variation law implied by Table 7 is quite different from that in Table 3 because it involves growth first, then a decrease. The global maximum is near the center inside the insulation spacer because it is nearest to the interface between the insulation spacer and the energized conductor. Thus, we can conclude that the maximum electric field intensity on the air gap surface is positively correlated with the distance from the charged interface. Furthermore, the observed surface electric field intensity is greater than 3 kV/mm, regardless of how the air gap is distributed, indicating an easily triggered partial discharge.

Table 7. Electric field distribution and maximum value on the air gap surface.

X-dis	15 mm	16 mm	17 mm	18 mm
E_{max}		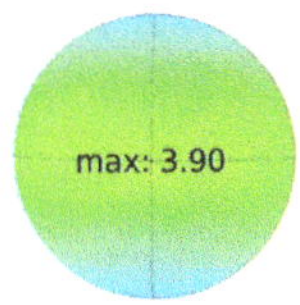	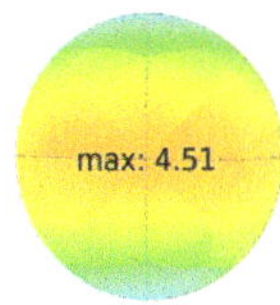	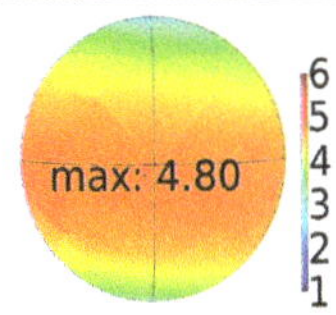
X-dis	19 mm	20 mm	21 mm	22 mm
E_{max}	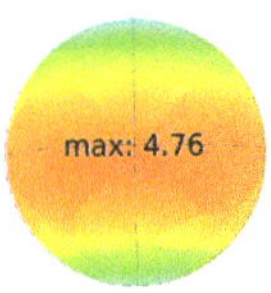	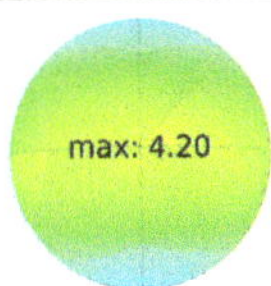	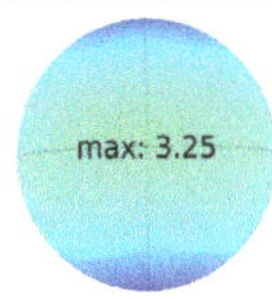	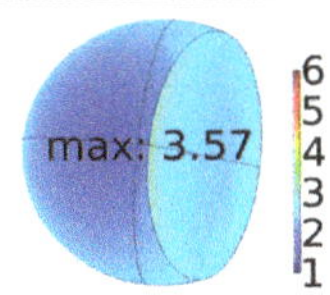

Table 7 suggests that the electric field intensity is not sharply affected by the appearance of the surface scratch but reveals a minimum value along the distribution pass. Figure 11 further shows the electric field intensity distribution at two surface scratches.

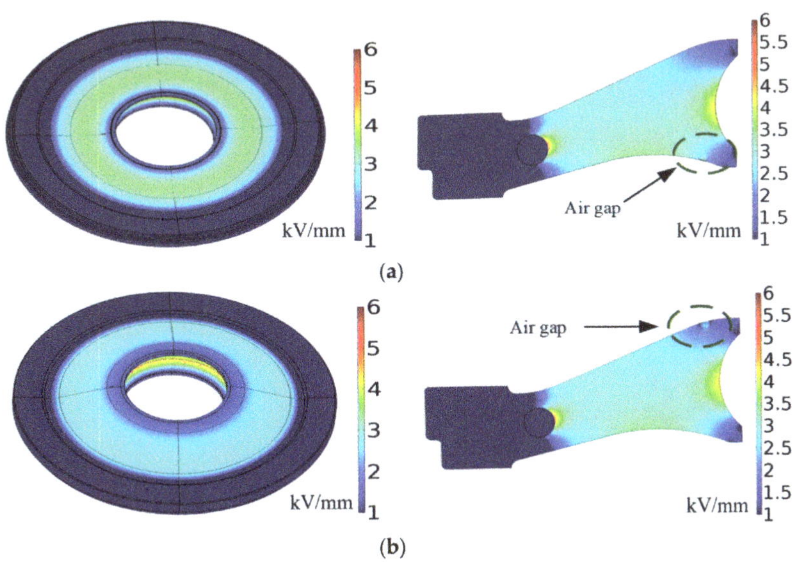

Figure 11. Electric field intensity distribution of a surface scratch at (**a**) 15 mm and (**b**) 20 mm.

The E_{max} values of an air gap corresponding to the upper and lower surfaces are compared for models with and without defects, similar to the analysis in Figure 7. The result differs from Figure 7 in that the electric field intensity of several points at the start or end position is lower than on the spacer surface. The distribution for the upper or lower surface of either the defect or defect-free model is not affected, which is the same as what is found in Figure 7. The relationship between E_{max} and axial distance is described via fitting analysis in Figure 12. Finally, Equation (6) is used to calculate the amplitude ratio of E_{max} corresponding to the air gap (E_{max_air}) and its upper or lower surface (E_{max_upper} or E_{max_lower}). These amplitude ratios are shown in Table 8. The biggest ratio is only 1.641 and much less than 1.999 in Table 4. Thus, the axially distributed air gap has less impact than the radial gap on the electric field intensity distribution.

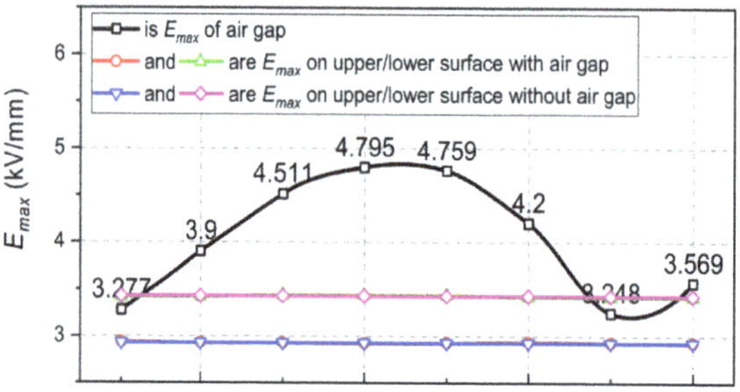

(a)

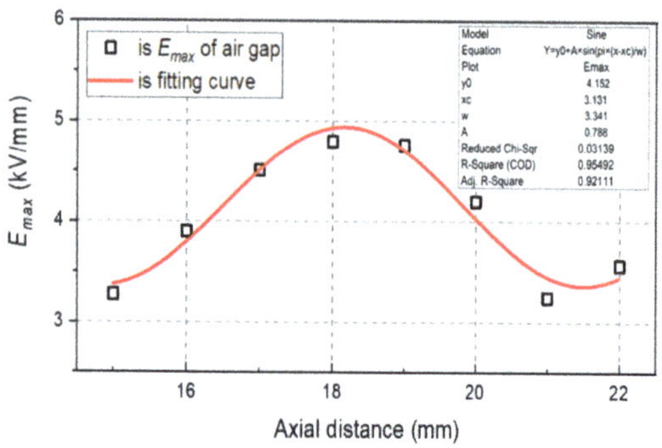

(b)

Figure 12. (**a**) Comparison of electric field intensities at different positions. (**b**) Fitting analysis of electric field intensity and radial distance.

Table 8. Amplitude ratio of electric field intensity of an air gap compared with the ratios for the upper and lower surfaces.

X-dis	15 mm	16 mm	17 mm	18 mm
k_{e_upper}	1.117	1.332	1.541	1.641
k_{e_lower}	0.959	1.141	1.314	1.397
X-dis	19 mm	20 mm	21 mm	22 mm
k_{e_upper}	1.626	1.431	1.109	1.220
k_{e_lower}	1.386	1.229	0.950	1.044

4.3. Impact on Mechanical Stress Distribution

The mechanical stress distribution of an air gap is shown in Table 9 for various axial distances. Another trend can be found when this is compared with the radial distribution in Table 5. According to Table 9, the stress amplitude of the air gap surface is negatively related to the distance from the pressure side. The lower surface is assigned to be the pressure side (with an atmospheric pressure of 2.4 MPa); thus, a smaller axial distance means higher mechanical stress is generated on the surface of the air gap. Although another surface scratch appears at an axial distance of 22 mm, it is located (placed) on the upper surface, which is far away from the pressure side, causing minimum mechanical stress. Figure 13 shows the stress distribution of surface scratches. In the picture, the maximum stress on the entire spacer is also near the interface between the insulation spacer and the metal shell.

Figure 14 presents the mentioned law of decrease in P_{max} versus axial distance. When an air defect is inserted inside the insulation spacer, the mechanical stress of the upper surface varies with the changing axial distance and shows no functional relationship. In contrast, a functional relationship between the P_{max} of the air gap and axial distance is determined by the fitting analysis in Figure 14b, which establishes a general decay tendency.

Table 9. Mechanical stress distribution and its maximum value on the air gap surface.

X-dis	15 mm	16 mm	17 mm	18 mm
P_{max}		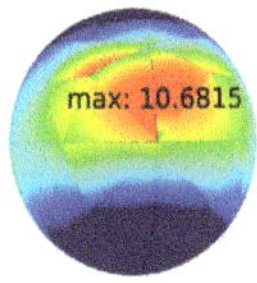	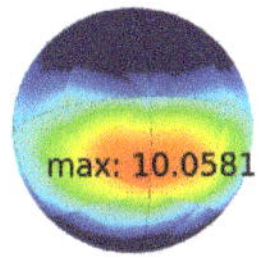	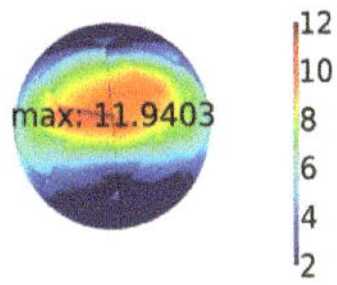
X-dis	**19 mm**	**20 mm**	**21 mm**	**22 mm**
P_{max}	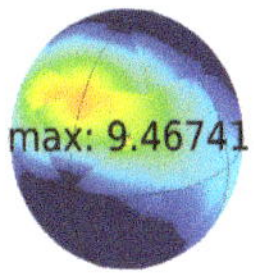			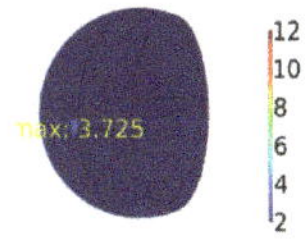

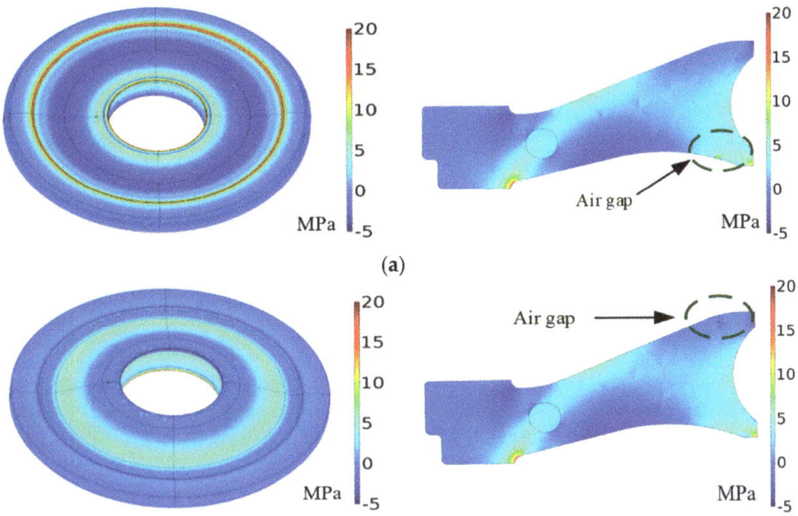

Figure 13. Stress distribution of a surface scratch at (**a**) 15 mm and (**b**) 20 mm.

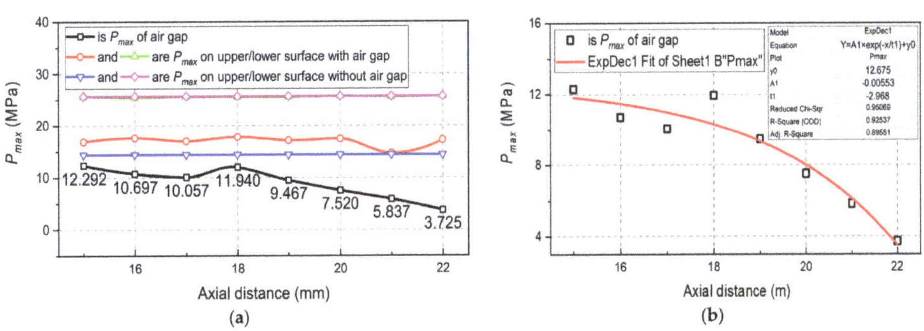

Figure 14. (**a**) Comparison of electric field intensities at different positions. (**b**) Fitting analysis of electric field intensity and radial distance.

The recorded P_{max} for the air gap at 18 mm is higher than at the last point (17 mm). This is attributed to the shorter distance from this point to the conductor than from other points. The same results are also obtained in Figure 9. Consequently, the stress on the air gap depends on the distance between the air gap and the pressure side or other interfaces (because these interfaces are also pressure boundaries and always sustain the local maximum of the mechanical stress). The pressure increases with decreasing distance.

Table 10 shows the amplitude ratio of P_{max} recorded for an air gap and upper and lower surfaces of the insulation spacer. The air gap defect only occurs on the lower surface of the insulation spacer, forming a surface scratch, and the mechanical stress at this time reaches half of that of the surface level. Thus, the air gap on the spacer surface leads to an obvious stress distortion, reducing the mechanical capacity of the defect insulation spacer by concentrating the surrounding stress at this point.

Table 10. Electric field distribution and its maximum value on the air gap surface.

X-dis	15 mm	16 mm	17 mm	18 mm
k_{e_upper}	0.729	0.608	0.592	0.672
k_{e_lower}	0.479	0.421	0.393	0.468
X-dis	19 mm	20 mm	21 mm	22 mm
k_{e_upper}	0.553	0.432	0.399	0.217
k_{e_lower}	0.371	0.293	0.229	0.145

5. Conclusions

A radially distributed air gap does not affect the electric field intensity and stress distribution of the upper and lower surfaces. The electric field intensity of the radially distributed air gap is 1.2–1.7 times the maximum value on the spacer surface. In contrast, the maximum stress on the air gap is only 0.08–0.77 times that on its surface. Therefore, the air gap significantly affects the surrounding electric field distribution. A significant air gap discharge occurs when the electric field intensity is greater than 3 kV/mm, causing a breakdown and partial discharge.

For an axial air gap distribution, the electric field strength of the air gap surface grows with increasing distance from the surface. The maximum value (4.795 kV/mm) occurs near the center conductor, and the local electric field strength is minimum on the outer surface. However, the maximum stress (12.292 MPa) of the air gap occurs on the lower surface. Consequently, the radial air gap distribution contributes more significantly to the electric field distribution, whereas the axial distribution dominates the distortion of the surface stress in a different way.

Author Contributions: Y.D. and X.F. designed the experiments and wrote the manuscript; X.F., F.L. and C.L. are all corresponding authors; all authors analyzed the data and approved the submitted manuscript. All authors have read and agreed to the published version of the manuscript.

Funding: This work was supported by State Key Laboratory of Power System Operation and Control, Tsinghua University (Grant No. SKLD22M03), Tsinghua-Jiangyin Innovation Special Fund (Grant No. 2022JYTH01), and Taikai Innovation Funding (Grant No. JTCB202209210002).

Data Availability Statement: The data supporting the findings of this study are available from the corresponding author upon reasonable request.

Conflicts of Interest: The authors declare no conflict of interest.

References

1. Li, C.; Zhang, C.; Lv, J.; Liang, F.; Liang, Z.; Fan, X.; Riechert, U.; Li, Z.; Liu, P.; Xue, J.; et al. China's 10-year progress in DC gas-insulated equipment: From basic research to industry perspective. *Ienergy* **2022**, *1*, 400–433. [CrossRef]
2. Fan, X.H.; Liu, J.F.; Goh, H.W.; Zhang, Y.; Wu, T.; Liu, J. Aging Analysis of Transformer Insulation at Weakest Region: Dielectric Parameters Extraction via Immune Optimization. *IEEE Trans. Transp. Electrif.* **2022**, *9*, 1579–1589. [CrossRef]

3. Fan, X.H.; Liu, J.F.; Goh, H.W.; Zhang, Y.; Zhang, G.; Rahman, S. Acquisition of FDS for Oil-Immersed Insulation at Transformer Hotspot Region Based on Multiconstraint NSGA Model. *IEEE Trans. Ind. Electrons.* **2022**, *69*, 13625–13635. [CrossRef]
4. Li, C.; Yang, Y.; Xu, G.; Zhou, Y.; Jia, M.; Zhong, S.; Gao, Y.; Park, C.; Liu, Q.; Wang, Y.; et al. Insulating materials for realising carbon neutrality: Opportunities, remaining issues and challenges. *High Volt.* **2022**, *7*, 610–632. [CrossRef]
5. Jia, C.; Zhao, W.; Zhu, Y.; Lu, W.; Tang, Z. A Numerical Study on the Decomposition and Diffusion Characteristics of SF6 in Gas-Insulated Switchgear with Consideration of the Temperature Rising Effect. *Energies* **2022**, *15*, 7834. [CrossRef]
6. Rachmawati; Kojima, H.; Kato, K.; Zebouchi, N.; Hayakawa, N. Electric Field Grading and Discharge Inception Voltage Improvement on HVDC GIS/GIL Spacer with Permittivity and Conductivity Graded Materials (ε/σ-FGM). *IEEE Trans. Dielectr. Electr. Insul.* **2022**, *29*, 1811–1817. [CrossRef]
7. Du, B.X.; Dong, J.N.; Liang, H.C. Electric field control by permittivity functionally graded and superficially non-linear conductivity materials for DC-GIS spacer. *High Volt.* **2022**, *7*, 992–1000. [CrossRef]
8. Zhao, C.; Tang, Z.; Zhang, L.; Wu, D.; Xia, H.; Li, H.; Pan, K. Entire process of surface discharge of GIS disc-spacers under constant AC voltage. *High Volt.* **2020**, *5*, 591–597. [CrossRef]
9. Yang, D.; Tang, J.; Zeng, F.; Yang, X.; Yao, Q.; Miao, Y.; Chen, L. Correlation characteristics between SF6 decomposition process and partial discharge quantity under negative DC condition initiated by free metal particle defect. *IEEE Trans. Dielectr. Electr. Insul.* **2018**, *25*, 574–583. [CrossRef]
10. Lu, L.; Ichimura, S.; Rokunohe, T. Interaction Between Partial Discharge and Generated Bubbles Under Repeated Lightning Impulses in Transformers Using a Complex Structure Model. *IEEE Trans. Dielectr. Electr. Insul.* **2021**, *28*, 727–735. [CrossRef]
11. Yadam, Y.R.; Sarathi, R.; Arunachalam, K. Numerical and Experimental Investigations on Influence of Internal Defect Parameters on Partial Discharge Induced UHF Signals in Gas Insulated Switchgear. *IEEE Access* **2022**, *10*, 110785–110795. [CrossRef]
12. Shiling, Z. Relationship between Electric Field Distribution Computation and Insulation Performance of GIS Spacer. In Proceedings of the 2020 IEEE 6th International Conference on Computer and Communications (ICCC), Chengdu, China, 11–14 December 2020; pp. 2397–2402.
13. Li, X.; Liu, W.; Xu, Y.; Ding, D. Discharge Characteristics and Detectability of Metal Particles on the Spacer Surface in Gas-Insulated Switchgears. *IEEE Trans. Power Del.* **2022**, *37*, 187–196. [CrossRef]
14. Wang, W.; Gao, Y.; Zhao, H. The Effect of a Metal Particle on Surface Charge Accumulation Behavior of Epoxy Insulator with Zoning Coating. *Energies* **2022**, *15*, 4730. [CrossRef]
15. Cho, H.G.; Hong, Y.-G.; Noh, Y.-H.; Bae, J.; Lee, J. Effect of Bio-based Epoxy Resins on the Thermal, Mechanical Properties and their Applicability to Gas Insulated Switchgear Spacer Materials. In Proceedings of the 2022 6th International Conference on Electric Power Equipment-Switching Technology (ICEPE-ST), Seoul, Republic of Korea, 15–18 March 2022; pp. 271–274.
16. Chen, T.; Shafai, C.; Rajapakse, A.; Liyanage, J.; Neusitzer, T. Micromachined AC/DC electric field sensor with modulated sensitivity. *Sens. Actuator A Phys.* **2016**, *245*, 76–84. [CrossRef]
17. Li, Z.; Yuan, H.; Cui, Y.; Ding, Z.; Zhao, L. Measurement of distorted power-frequency electric field with integrated optical sensor. *IEEE Trans. Instrum. Meas.* **2019**, *68*, 1132–1139. [CrossRef]
18. Ji, S.; Jia, Y.; Yang, X.; Liu, Z.; Zhu, L. A Method for Measuring Electric Field Distribution along Insulator Surfaces Based on Electroluminescence Effect and Its Image Processing Algorithm. *IEEE Trans. Dielectr. Electr. Insul.* **2022**, *29*, 939–947. [CrossRef]
19. Moradnouri, A.; Vakilian, M.; Hekmati, A.; Fardmanesh, M. The End Part of cryogenic HV Bushing Insulation Design in a 230/20 kV HTS Transformer. *Cryogenics* **2020**, *108*, 103090. [CrossRef]
20. Ji, S.; Jia, Y.; Yang, X.; Liu, Z.; Zhu, L. Simulation and experimental study on mechanical performance of UHV GIS spacer in hydrostatic test. *IEEE Trans. Dielectr. Electr. Insul.* **2017**, *24*, 3122–3313.
21. Wang, Q.; Fan, X.; Ding, Z.; Hwang, G.H.; Zhang, Y.; Liu, J. A Genetic Optimization Model for Aging Analysis of Paper Insulation in High-Temperature Region via Frequency Domain Spectroscopy. *IEEE Trans. Transport. Electrific.* **2022**. [CrossRef]
22. Talaat, M.; El-Zein, A.; Amin, M. Electric field simulation for uniform and FGM cone type spacer with adhering spherical conducting particle in GIS. *IEEE Trans. Dielectr. Electr. Insul.* **2018**, *25*, 339–351. [CrossRef]
23. Wang, H.; Peng, Z.; Guo, Z.; Wang, Q.; Liao, J.; Zhang, S. Simulation study of electric influence caused by defects on UHV AC GIS spacer. In Proceedings of the 2015 IEEE 11th International Conference on the Properties and Applications of Dielectric Materials (ICPADM), Sydney, NSW, Australia, 19–22 July 2015; pp. 676–679.
24. Moradnouri, A.; Vakilian, N.; Hekmati, A.; Fardmanesh, M. HTS Transformer's Partial Discharges Raised by Floating Particles and Nitrogen Bubbles. *J. Supercond. Nov. Magn.* **2020**, *33*, 3027–3034. [CrossRef]
25. Liang, Z.; Lin, C.; Liang, F.; Zhuang, W.; Xu, Y.; Tang, L.; Zeng, Y.; Hu, J.; Zhang, B.; Li, C.; et al. Designing HVDC GIS/GIL spacer to suppress charge accumulation. *High Volt.* **2022**, *7*, 645–651. [CrossRef]
26. Li, X.; Wan, M.; Zhang, G.; Lin, X. Surface Charge Characteristics of DC-GIL Insulator under Multiphysics Coupled Field: Effects of Ambient Temperature, Load Current, and Gas Pressure. *IEEE Trans. Dielectr. Electr. Insul.* **2022**, *29*, 1530–1539. [CrossRef]

energies

MDPI

Article

Composite Insulator Defect Identification Method Based on Acoustic–Electric Feature Fusion and MMSAE Network

Bizhen Zhang [1], Shengwen Shu [1,*], Cheng Chen [2], Xiaojie Wang [3], Jun Xu [3] and Chaoying Fang [3]

[1] School of Electrical Engineering and Automation, Fuzhou University, Fuzhou 350108, China; 210127165@fzu.edu.cn
[2] Fuzhou Power Supply Company, State Grid Fujian Electric Power Co., Ltd., Fuzhou 350009, China; chencheng938315@gmail.com
[3] Electric Power Research Institute, State Grid Fujian Electric Power Co., Ltd., Fuzhou 350007, China; wang_xiaojie99@163.com (X.W.); xujun8127@163.com (J.X.); fc5658779@163.com (C.F.)
* Correspondence: shushengwen@fzu.edu.cn

Abstract: Aiming to solve the partial discharge problem caused by defects in composite insulators, most existing live detection methods are limited by the subjectivity of human judgment, the difficulty of effective quantification, and the use of a single detection method. Therefore, a composite insulator defect diagnosis model based on acoustic–electric feature fusion and a multi-scale perception multi-input of stacked auto-encoder (MMSAE) network is proposed in this paper. Initially, during the withstanding voltage experiment, the electromagnetic wave spectrometer and ultrasonic detector were used to collect and process the data of six types of composite insulator samples with artificial defects. The electromagnetic wave spectrum, ultrasonic power spectral density, and *n-S* map were then obtained. Then, the network architecture of MMSAE was built by integrating a stacked auto-encoder and multi-scale perception module; the feature extraction and fusion methods of the electromagnetic wave spectrum and ultrasonic signal were investigated. The proposed method was used to diagnose test samples, and the diagnostic results were compared to those obtained using a single input source and the artificial neural network (ANN) method. The results demonstrate that the detection accuracy of acoustic–electric feature fusion is greater than that of a single feature; the accuracy of the proposed method is 99.17%, which is significantly higher than the accuracy of the conventional ANN method. Finally, composite insulator defect diagnosis software based on PYQT5 and Keras was developed. Ten 500 kV aging composite insulators were used to validate the effectiveness of the proposed method and design software.

Keywords: composite insulator; defect identification; deep learning; feature fusion; electromagnetic wave spectrum; ultrasonic

1. Introduction

Composite insulators are widely used in China due to their light weight, high strength, good resistance to pollution, and ease of operation and maintenance [1,2]. However, in the long-term operation process, the composite insulator will experience issues such as hydrophobicity decrease, mechanical strength decrease, insulation strength decrease, shed sheath fall, core rod fracture and degradation under the combined action of high field strength, high mechanical stress, air pollution, and other factors, which pose a threat to the safe operation of the power grid [3]. Under the influence of a strong electric field, partial discharge will occur when there are defects within the composite insulator, resulting in several physical and chemical effects, including an increase in leakage current, electric field distortion, electromagnetic wave, and ultrasonic radiation [4–7]. As a result, it is critical to investigate an effective composite insulator defect diagnosis method to improve operational reliability.

To solve the problem of composite insulator defect detection in operation, various detection methods have been studied at home and abroad [8–11]. In [12], the phase resolved partial discharge (PRPD) spectrum is used to determine the operating state of an insulator using the pulse current method. In [13], the 3D electric field of an insulator was simulated using finite element software, and the effect of surrounding air conductivity on the electric field was evaluated. In [14], an electric field sensor based on the Pockels effect was utilized to construct an insulator defect detection platform. In [15], the PD signals of six artificial insulation defect models were measured using the UHF method, which achieved the same accuracy as ANN and SVM, had a faster recognition speed, and was robust with respect to certain samples. A non-destructive ultrasonic phased array technology was utilized in [16] to detect defective silicone rubber composite insulator samples. In [17], an electrical equipment insulation defect detection system based on infrared and ultraviolet optoelectronics-sensing technology was developed. The insulation condition of the electrical equipment was evaluated comprehensively using the signals of infrared and ultraviolet optoelectronics sensors. However, most detection methods have drawbacks, such as expensive equipment and environmental sensitivity.

Pattern recognition is the study of how machines observe the environment, learn to extract useful features from the environment, and classify patterns rationally [18–21]. In [22], K-means clustering was applied to the PRPD spectrum, and different types of discharges were distinguished through feature extraction and data processing. Based on the support vector machine and probabilistic neural network, the authors of [23] extracted the probability density function of the mean, standard deviation, peak, and skewness of the wavelet packet coefficients by wavelet packet decomposition of PD signals and confirmed that the support vector machine is more precise. In [24], the wavelet packet decomposition of the PD signal was used to extract features, and the classification and regression tree algorithm, the bagging algorithm, and the regression random forest were used to locate the PD. In [25], a partial discharge pulse pattern recognition method based on a fuzzy decision tree was proposed. First, the C4.5 algorithm was used to extract fuzzy rules from the data, followed by the backpropagation training algorithm to fine-tune the fuzzy rules' parameters. The authors of [26] proposed a CWD time–frequency spectrum based on variational mode decomposition and an optimized convolutional neural network with cross-layer feature fusion. In [27], a feature map produced by variational mode decomposition and the stacked sparse auto-encoder network were combined to determine the four discharge types. The authors of [28] proposed a method for separating and classifying pulse-shaped signals based on a fuzzy classifier that can identify and separate multiple PD sources. However, most current defect-recognition algorithms rely on a single physical signal, which is insufficient for a comprehensive evaluation of the detected target's condition. The application of deep learning can remedy this deficiency.

The gaps left by previous studies to be filled include: (1) most of the detection methods of composite insulator defects have the subjectivity of human judgment, difficulty in effective quantification, and limitation of a single detection method; and (2) the composite insulator defect-recognition algorithm is mostly based on a single physical signal and various detection methods cannot achieve information sharing.

This paper proposes a composite insulator defect diagnosis model based on acoustic–electric feature fusion and deep learning. The main contributions of this paper are as follows:

1. Electromagnetic wave and ultrasonic live detection tests of composite insulator defects are carried out. Six typical defect insulator samples are made in the laboratory, and a defect detection test platform is built. The electromagnetic wave spectrometer and ultrasonic detector are used to measure the electromagnetic wave and ultrasonic signal radiated by the unpressurized background noise, the intact insulator and the defective insulator in a certain distance. FFT is used to analyze the spectral characteristics of electromagnetic wave signals, and the ultrasonic signal is analyzed by a power spectrum and *n*-*S* map. Combining electromagnetic wave and ultrasonic

detection methods can scientifically and effectively reflect different types of insulator defects.

2. A composite insulator defect diagnosis algorithm based on acoustic–electric feature fusion and multi-scale perception multi-input of a stacked auto-encoder network is proposed. The peak and average value spectrum of electromagnetic waves, an ultrasonic power spectrum and *n*-*S* map are extracted by a stacked auto-encoder and are reduced to similar dimensions to avoid one signal occupying too much weight and weakening the influence of other signals on the results. The above data are fused as the input of the multi-scale perception multi-input network. By stacking multiple such structures, data information can be extracted on multiple scales without increasing the number of network parameters, avoiding overfitting caused by too many network layers and improving the accuracy of classification. By comparison, it is proven that the detection accuracy of the proposed method is obviously better than that of a single-feature input source and traditional ANN method.
3. Composite insulator defect diagnosis software is developed, and the aging composite insulators replaced on-site are tested and verified. Based on the proposed acoustic–electric feature fusion and MMSAE network, a composite insulator defect diagnosis software is developed in PYQT5 and Keras environments. The software detection and dissection analysis of ten 500 kV aging composite insulators replaced on-site are carried out, which proves the accuracy of the proposed method and software. It is helpful to comprehensively study and judge the operation state and defect type of composite insulators on transmission lines, and this has practical engineering significance.

The remainder of the article is organized as follows. Section 2 describes the acoustic–electric feature test and results analysis of composite insulator defects. Section 3 describes the construction and analysis of the composite insulator defect identification model based on acoustic–electric feature fusion and the MMSAE network. In Section 4, examples and analyses of engineering applications are provided. Finally, the conclusions are given in Section 5.

2. Acoustic–Electric Features Test of Composite Insulator Defects

2.1. Test Circuit

Figure 1 depicts the test circuit diagram for the composite insulator defect acoustic–electrical feature test. The test circuit contains a power frequency test transformer, a composite insulator short sample, a protective resistance, and a ground electrode. The test insulator samples in this paper are all 1/10 short samples of DC $\pm$ 1100 kV, and the rated voltage is DC $\pm$ 110 kV. Therefore, under AC conditions, the peak voltage of the test insulator is set to 110 kV, and the effective value of the AC voltage applied to the test insulator by the test transformer is 77.8 kV.

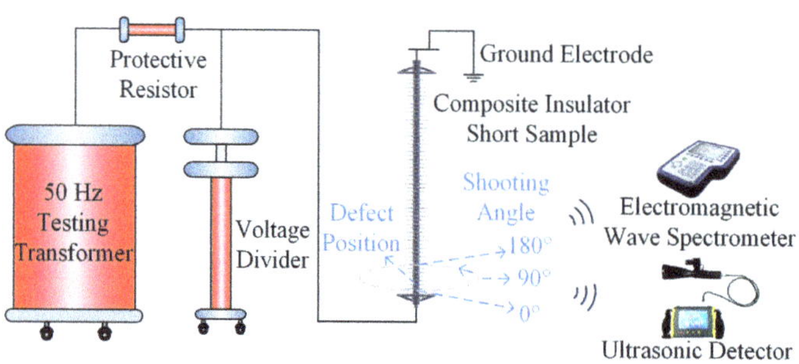

Figure 1. Test circuit.

During the test, data were collected from the front (0°), side (90°), and back (180°) of the defect, depending on the defect's orientation. The data acquisition position is 3~4 m away from the measured insulator. The electromagnetic wave and ultrasonic data of background noise without pressurization, intact insulator, and various simulated degraded insulators are measured, respectively.

2.2. Composite Insulator Defect Simulation

The primary cause of the failure of composite insulators is the uneven field strength of the interface, which is a result of defects and an imperfect manufacturing process, resulting in long-term carbonization channels and gradually developing into interface breakdown discharge. Therefore, by embedding copper wire or a carbon rod between the high-voltage end metal fittings and the first umbrella skirt to simulate conductive or semi-conductive defects, the defects and high-voltage end metal fittings are equipotential to simulate conduction-type defects and non-equipotential to simulate suspension-type defects. Separating the interlayer between the umbrella skirt sheath and the core rod creates an air gap space, which is then filled with a salt solution to simulate water vapor entering the interlayer. By smearing carbon powder on the surface of the high-voltage umbrella skirt, the leakage tracking defect caused by a flashover on the surface of the umbrella skirt can be simulated. According to the failure mechanism of insulators [29–32], the following six kinds of defects are simulated, as shown in Table 1. Figure 2 depicts the actual simulation of various defects. The copper wire and carbon rod are embedded in the core rod and sheath interface, each measuring 3 mm in diameter and 10 cm in length, with the ends being sharply polished. The carbon powder weighs 5 g, and the coating is uniform.

Table 1. Types and methods of defect simulation.

Defect Type	Starting Position	Simulated Material
Intact		
Conduction conductive Conduction semi-conductive Water vapor entry	High-voltage end metal fittings	Copper wire Carbon rod Salt solution
Suspension conductive Suspension semi-conductive Leakage tracking	High-voltage end first umbrella skirt	Copper wire Carbon rod Carbon powder

2.3. Experiment Results and Analysis

2.3.1. Processing Method and Result Analysis of Electromagnetic Wave Data

The PDS100 electromagnetic wave spectrometer is used to record the RF signal emitted by partial discharge. During gating time, the signal within the center frequency's bandwidth is detected, and the bandwidth of each band was 1.9 MHz. On the signals in the frequency band corresponding to each center frequency, an FFT transform is performed, and the spectrums corresponding to each frequency band are merged to form a complete spectrum. The experiment adopted S.P.A.M mode, which simultaneously records the spectrum's peak and average values.

The frequency components of 50 MHz~1 GHz include a variety of continuous signals, such as AM radio, mobile phone, and other communication signals. The background wave without voltage is first recorded, then the background noise is subtracted from the spectrum with voltage, and the frequency band amplitude of less than 0 is set to 0. Each spectrum contains 501 detection center frequencies ranging from 50 MHz to 1 GHz. Figure 3 depicts the electromagnetic wave spectrum results for each composite insulator defect.

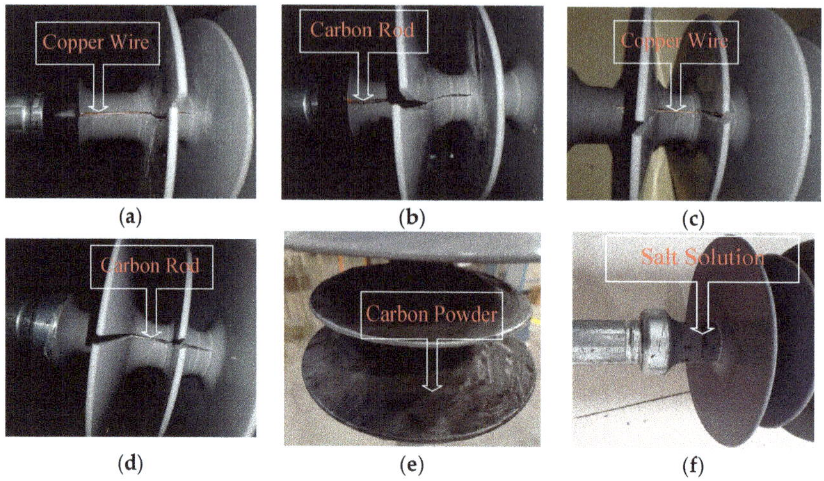

Figure 2. Actual simulation of various defects: (**a**) conduction conductive; (**b**) conduction semi-conductive; (**c**) suspension conductive; (**d**) suspension semi-conductive; (**e**) leakage tracking; (**f**) water vapor entry.

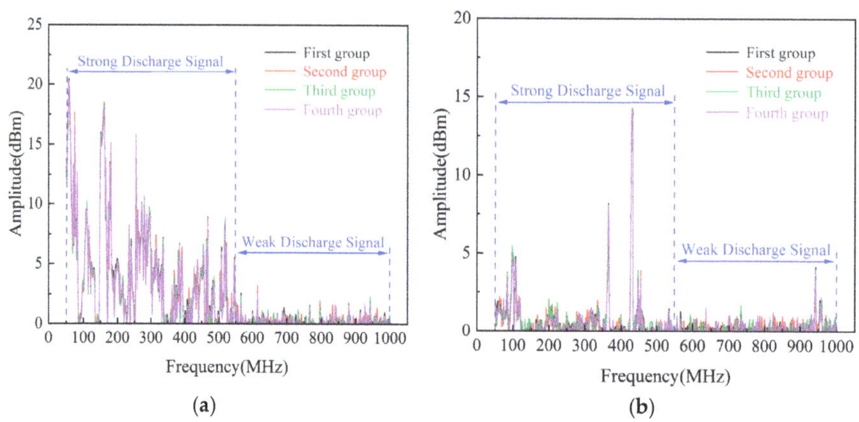

Figure 3. Four groups of electromagnetic wave spectrums of conduction conductive defect composite insulators: (**a**) peak value; (**b**) average value.

In this paper, four groups of electromagnetic wave characteristic detection experiments are carried out. The electromagnetic wave experimental results of conduction conductive composite insulators are shown in Figure 3. In the four groups of experiments, the peak value curve and the average value curve of the electromagnetic wave spectrum of the conduction conductive composite insulator have the same change trend. The situation of intact and other defect types of composite insulators is consistent with that of conduction conductive composite insulators. Therefore, in this paper, the electromagnetic wave spectrum of the first group of experiments is taken as an example to summarize the characteristics of intact and various types of defective composite insulators, as shown in Figure 4.

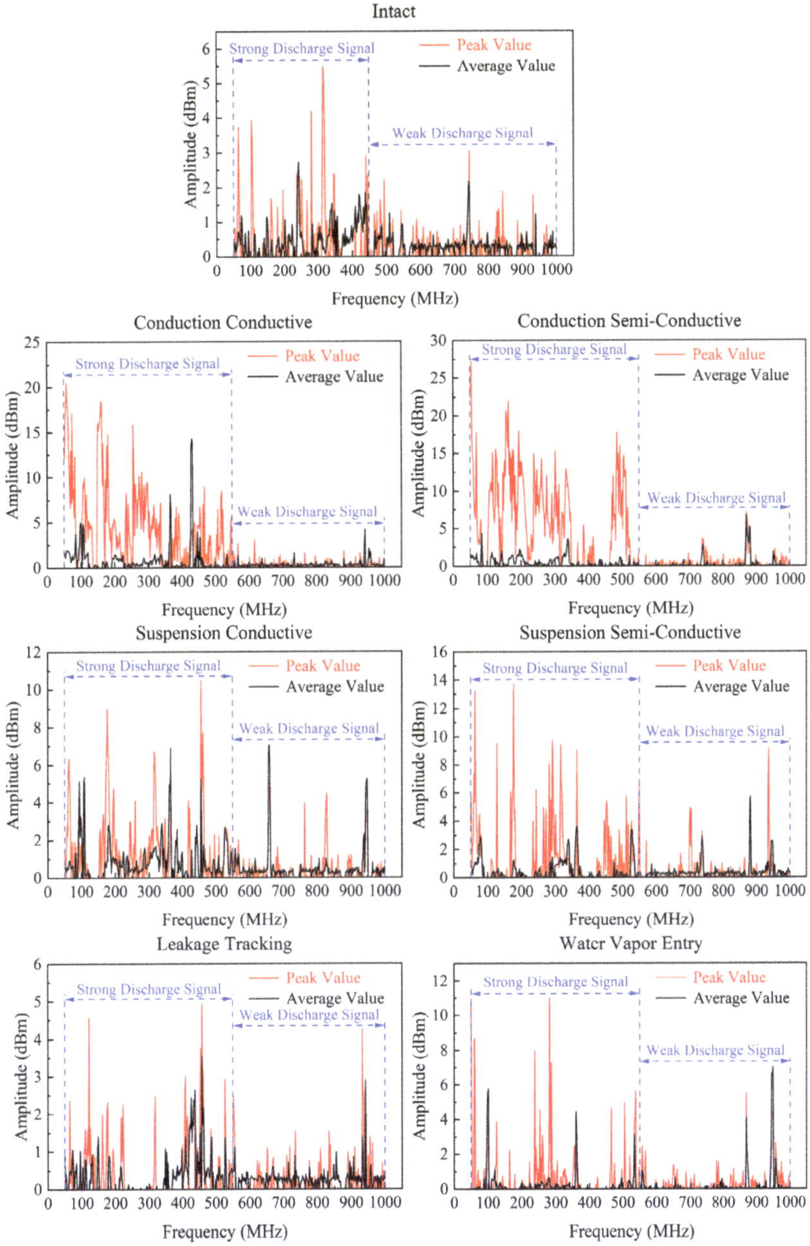

Figure 4. Electromagnetic wave spectrum of various insulator defects.

Figure 4 reveals that the intact insulator's electromagnetic wave spectrum contains a discharge signal at 0~450 MHz, indicating that this is the electrical or corona signal of the line or metal fittings. Comparing the electromagnetic wave spectrum of intact insulators with each defect spectrum reveals the following characteristics.

1. The peak spectrum and average spectrum of conduction defects have higher amplitude at 0~80, 120~200, and 450~550 MHz. The peak spectrum of 350~450 MHz is similar to that of an intact insulator, whereas the average spectrum is distinct.
2. The peak spectrum and average spectrum of suspension defects in the range 0~80 MHz are not as high as those of conduction defects, but they are still relatively obvious. The peak spectrum and average spectrum of 150~200 MHz have higher signals. The peak spectrum of 350~450 MHz and the average spectrum of 150~400 MHz are obviously different from the intact insulator signal.
3. Since there is little difference between the two types of frequency spectrums of the leakage tracking defect and that of the intact insulator, it is assumed that there is no obvious discharge in the defect.
4. The average spectrum and peak spectrum of the water vapor entry defect exhibit intermittent discharge peaks at 40, 160, 360, and 450~550 MHz. The average spectrum generally has a richer signal distribution and higher amplitude in the 50~600 MHz band.

2.3.2. Processing Method and Result Analysis of Ultrasonic Data

Figure 5 shows the specific flow of the processing method and result analysis of ultrasonic data. The SUD-300 ultrasonic detector is able to record ultrasonic signals radiated by partial discharges with a frequency greater than 20 kHz. Because the ultrasonic signal generated by partial discharge contains a wealth of signals at approximately 40 kHz, and because the frequency band can effectively eliminate the interference signal, the detection center frequency of the detector is selected at 40 kHz. Through the ultrasonic detector, the signal is converted to a range of audible frequencies between 0 and 4 kHz.

The background noise of three locations devoid of voltage, the ultrasonic shapes of intact insulators under a working phase voltage of 77.8 kV, and the ultrasonic shapes of various fault-simulated insulators in the frequency range of 38~42 kHz are respectively measured. The sampling frequency is 8 kHz, and the duration of each recording is 10 s.

(1) Wavelet packet denoising.

Since the experimental data acquisition in this paper was conducted in a laboratory, it can be assumed that the data are nearly free of noise. However, ultrasonic equipment operating in the field will be subject to various types of field noise, including wire corona signal interference and instrument internal interference. In addition, in long-distance fault detection, due to the scattering and attenuation characteristics of ultrasonic waves in the measured object and air, the received signals cannot meet the application requirements. Therefore, the collected signal must be denoised. The method of wavelet packet transform threshold denoising is utilized in this paper.

White noise and background noise are superimposed with the original signal at signal-to-noise ratios of 2, 5, 10, and 20 dB to test the filtering effect of the wavelet packet denoising method. After many aspects of comparison, the final selection consists of soft threshold function, unbiased likelihood estimation rules, wavelet function bior 2.4, and decomposition layer 4.

(2) Improved *n*-*S* map based on the PRPD method.

This paper proposes a general *n*-*S* map method suitable for ultrasonic testing under AC and DC conditions, based on the PRPD method. In the pulse current method, due to the absence of phase information in the DC discharge, only the number of discharges n and the apparent discharge q are selected as the pulse current characteristic quantity. Let n represent the number of zero crossing points of the ultrasonic shape in the ultrasonic method. Then, n can be analogous to the number of discharges in PRPD. Between the two adjacent zeros, the area S bounded by the acoustic wave and the abscissa axis can be analogous to the discharge amount in PRPD, and the area of all packages can be determined. The frequency distribution histogram *n*-*S* spectrum and box diagram of different defects can be drawn by statistical method. On this basis, various characteristic parameters are extracted for the classifier's input.

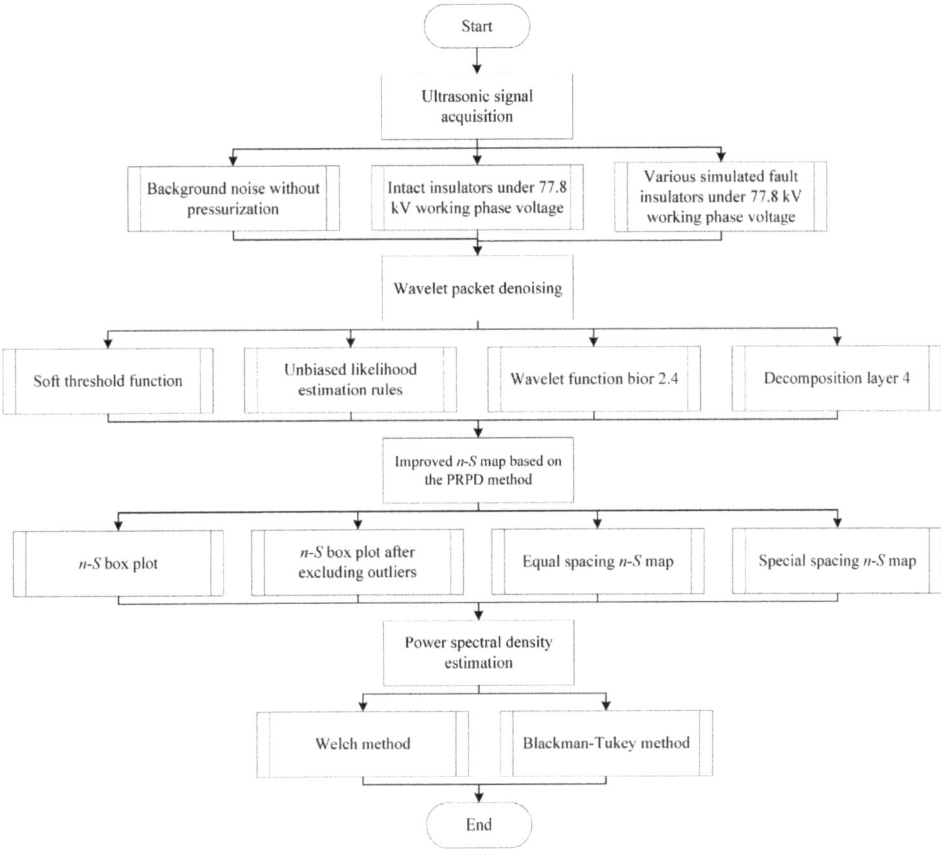

Figure 5. Flow of processing method and result analysis of ultrasonic data.

A box plot is a statistical graph used to display a set of scattered data. Box plot intuitively displays, from a statistical point of view, distribute characteristics of discrete data and are unaffected by outliers. They can also be used to clean data.

All enveloped areas S of the acoustic wave and the abscissa axis are calculated using the above method, and K at the upper and lower edges is 1.5. The data outside of this range are labeled as abnormal, and the statistical relationship between the number of zero crossings, n, and S is plotted, as shown in Figure 6. In the case of intact and leakage tracking, the position and median of the box are close to zero, and there is almost no discharge event, as shown in Figure 6. Water vapor entry defects discharge minimally, conduction defects discharge severely, and suspension defects discharge with lesser intensity than conduction defects.

Observing the received signal strength from various directions, the signal strength and density are greatest in front of the defect. This is due to the fact that the ultrasonic signal is received directly by the ultrasonic sensor through the air, rather than being blocked by the core rod sheath. The signal strength on the back of the defect is typically weak, especially for semi-conductive defects, which may be due to the fact that the core rod separates the direction of the received signal from the signal source and attenuates the ultrasonic signal. The outliers marked by red dots were removed and the box plot was redrawn, as shown in Figure 7.

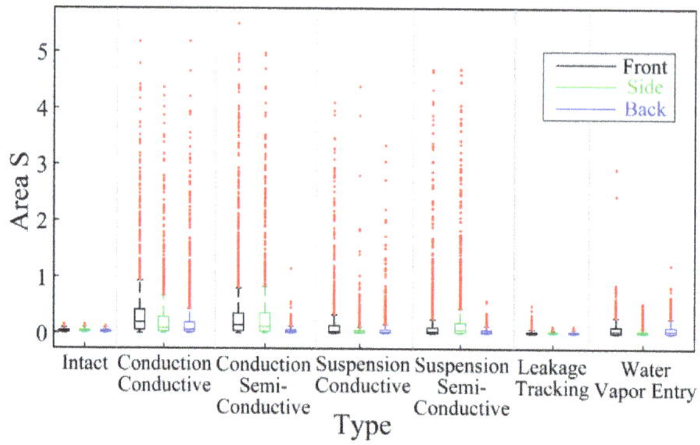

Figure 6. *n*-*S* box plot.

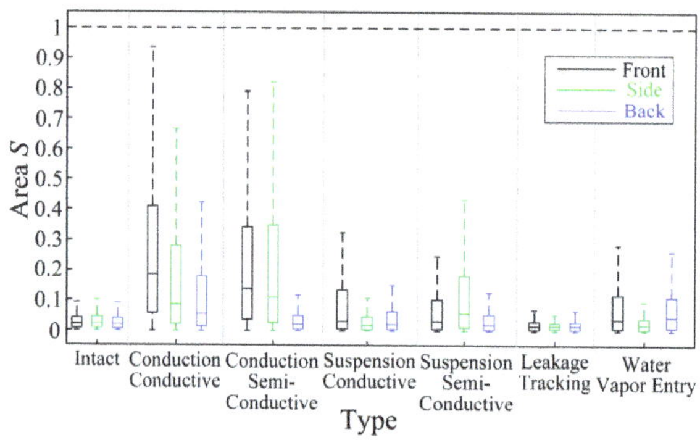

Figure 7. *n*-*S* box plot after excluding outliers.

To compare the *n*-*S* map of different defects under the same standard after excluding the outliers of *S* due to the large differences in amplitude and number of different types of discharges, the upper limit of the range of *S* is set as the global maximum, and all *S* are normalized to the range of (0, 1). As shown in Figure 8, this range was divided into 200 equal parts, and the *n*-*S* map was drawn with equal-spacing statistics.

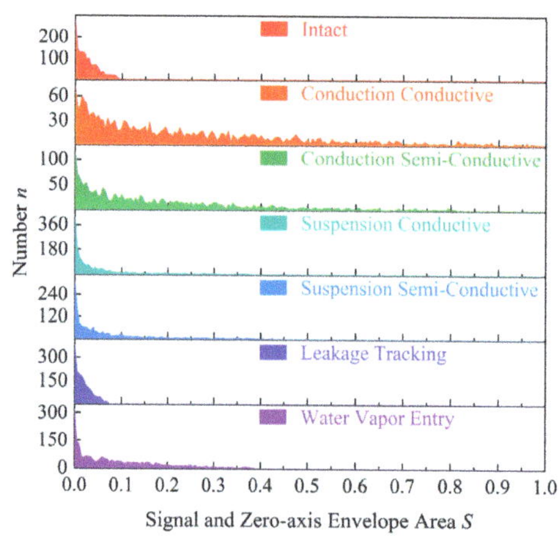

Figure 8. Equal-spacing *n*-*S* map.

As can be seen from the above figure, compared with the maximum envelope area S: conduction conductive > conduction semi-conductive > suspension conductive ≈ suspension semi-conductive ≈ water vapor entry > leakage tracking ≈ intact. From the perspective of the number n, the number n of suspension defects, leakage tracking defects, and water vapor entry defects in the small S region are relatively high, the number of conduction defects is relatively low, and the number of intact insulators in all S regions is low, confirming that leakage tracking defects and other defects are weak with discharge, but the number of discharges is greater.

The degree of data distribution skewness can be measured by the calculation of *variance*, *skewness* and *kurtosis*.

Variance is used to measure the degree of deviation between a random variable or a set of data and its mathematical expectation. The formula is as follow:

$$\sigma(X) = E\left\{\sum[X - \mu(X)]^2\right\} \tag{1}$$

where X is the statistical sample, μ is the sample mean, σ is the sample variance, and E represents the mean. *Variance* can be used to accurately describe the degree of deviation between the statistical data and the mean.

Skewness is a measure of the direction and degree of skew of statistical data distribution, which is used to measure the asymmetry of statistical data distribution. It can be calculated as follows:

$$Skew(X) = E\left[\left(\frac{X-\mu}{\sigma}\right)^3\right] \tag{2}$$

In the normal distribution *Skewness* = 0, positive bias distribution *Skewness* > 0, negative bias distribution *Skewness* < 0.

Kurtosis is also called the *kurtosis* coefficient. It is used to characterize the concentration of statistical data distribution near the mode. Intuitively reflected in the sharpness of the curve peak, the formula is as follows:

$$Kurt(X) = E\left[\left(\frac{X-\mu}{\sigma}\right)^4\right] \tag{3}$$

Normal distribution *Kurt* = 3, in the case of the same σ, *Kurt* > 3 is the leptokurtic state, corresponding to a thick tail; *Kurt* < 3 is the platykurtic state, corresponding to a thin tail. Table 2 displays the results of calculating the *skewness* and *kurtosis* of the envelope area S distribution for the seven types of ultrasonic signals listed above.

Table 2. The skewness and kurtosis of data S.

Area S	*Variance*	*Skewness*	*Kurtosis*
Intact	0.0005	0.90	3.10
Conduction conductive	0.0561	1.14	3.55
Conduction semi-conductive	0.0362	1.27	3.97
Suspension conductive	0.0076	1.85	5.72
Suspension semi-conductive	0.0100	1.61	4.77
Leakage tracking	0.0003	0.92	3.13
Water vapor entry	0.0092	1.04	3.26

The frequency distribution of area S exhibits an evidently positive skewness distribution close to an exponential distribution and *Skewness* > 0, *Kurtosis* > 3. Consequently, the size of the adjacent interval of the frequency distribution histogram is selected in a specific manner:

$$x_i = \frac{a^i - 1}{a^{N+1} - 1}\ (i = 0, 1, 2, \ldots, N+1) \quad (4)$$

where x_i is the selected abscissa point, [0,1] is the value range, and N is the number of statistical histograms or the number of regions divided into S. The unequal interval statistical method enables different defects to achieve good discrimination in areas with dense S distribution and to expand the statistical range in areas with sparse S distribution, thereby minimizing redundant data information. The value range of parameter a is (0,1) ∪ (1,+∞), which is related to the distribution pattern of data. When *Skewness* > 0, $a > 0$ is taken; when *Skewness* < 0, $0 < a < 1$ is taken, and the specific value is determined by specific needs.

In this paper, the n-S map is drawn by special spacing statistics, as shown in Figure 9, where $a = 1.03$, $N = 200$. The abscissa represents the statistical range sequence of the area S, that is, the number of intervals in the series $[x_i, x_{i+1}]$. The n-S map of special spacing demonstrates that special spacing can effectively avoid the situation in which data concentration in a particular region leads to low regional resolution. Consequently, it also has a good distinction effect on the skewness distribution, allowing it to more accurately reflect the distinction between various defects in the ultrasonic signal.

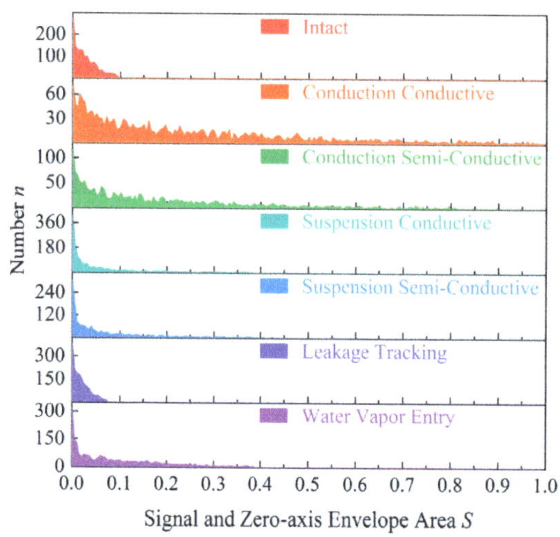

Figure 9. Special spacing n-S map.

From the distribution range of S, conduction defect > suspension defect > water vapor entry > leakage tracking > intact, which verifies the intensity of different defect discharge intensities. According to the number n, the conduction and water vapor entry defects have greater values in the region with a higher S sequence, the numerical fluctuation of suspension defects in the entire sequence region is small, and the n of the intact insulator and leakage tracking defects are predominantly distributed in the region with lower S sequence.

(3) Power spectral density estimation.

The power spectral density function, also known as the power spectrum, is a function that represents the variation of signal power with frequency in each unit frequency band, or the signal power distribution in the frequency domain. The Blackman–Tukey method and Welch method are used to generate two power spectra. Compared to the Blackman–Tukey method, the spectral estimation of the Welch method is smoother, and the selection of the window function reduces the possibility of modal mixing, thereby improving the resolution of power density spectral estimation.

The following frequencies are based on the actual frequency of the converted signal because the ultrasonic instrument converts the ultrasonic signal to the human ear to identify the acoustic wave, which is equivalent to reducing the frequency of the acoustic wave from 18~22 kHz to 0~4 kHz. All power spectral density estimation results are shown in Figure 10.

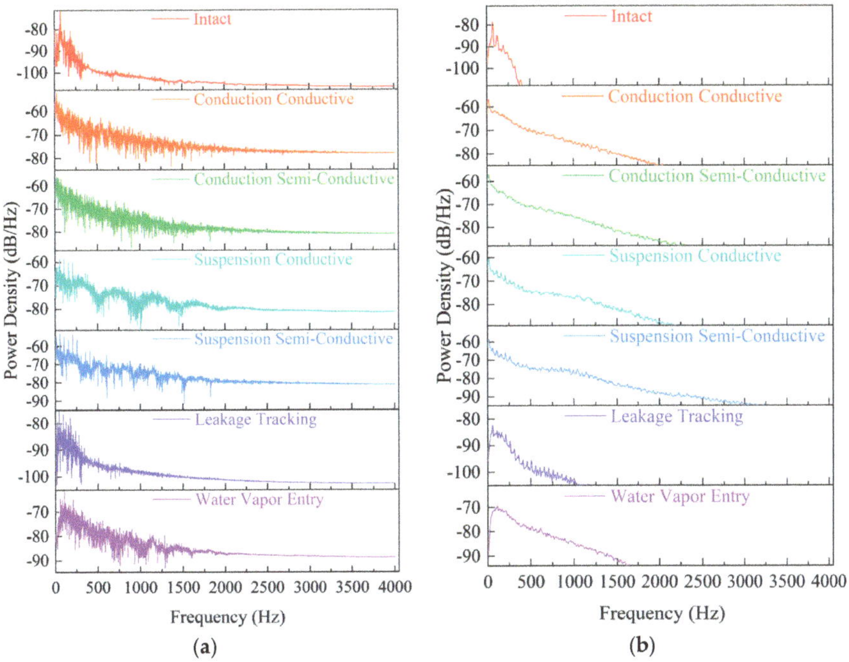

Figure 10. Power spectral density estimation: (**a**) Blackman–Tukey method; (**b**) Welch method.

Comparing the power spectral density estimations of the Blackman–Tukey method and the Welch method reveals that their amplitudes and graphs are similar. The Welch method is smoother, which is the result of piecewise averaging. Compared to the Blackman–Tukey method, the Welch method suppresses noise but also loses some signal detail components.

Comparing different defect signals, all exhibit the shape of high before and low after, with the maximum power density of the conductive defect being the highest, reaching −60 dB, the amplitude of the suspension defect being slightly lower, followed by water vapor entry, and the maximum power density of the intact insulator and the leakage tracking being the smallest, approximately −80 dB, which can be interpreted as signal interference caused by corona. Intact insulators, leakage tracking, and water vapor entry defects exhibit a partial rising edge in the frequency range of 0 to 100 Hz.

3. Composite Insulator Defect Identification Method Based on Acoustic–Electric Feature Fusion and MMSAE Network

3.1. Overall Network Architecture

This paper presents a method for diagnosing composite insulator defects based on acoustic–electric feature fusion and multi-scale perception multi-input of a stacked auto-encoder network. Figure 11 depicts the overall network architecture and feature dimensions of the method. The multi-input feature data are combined via their respective feature extraction layers and are fed into the classification layer.

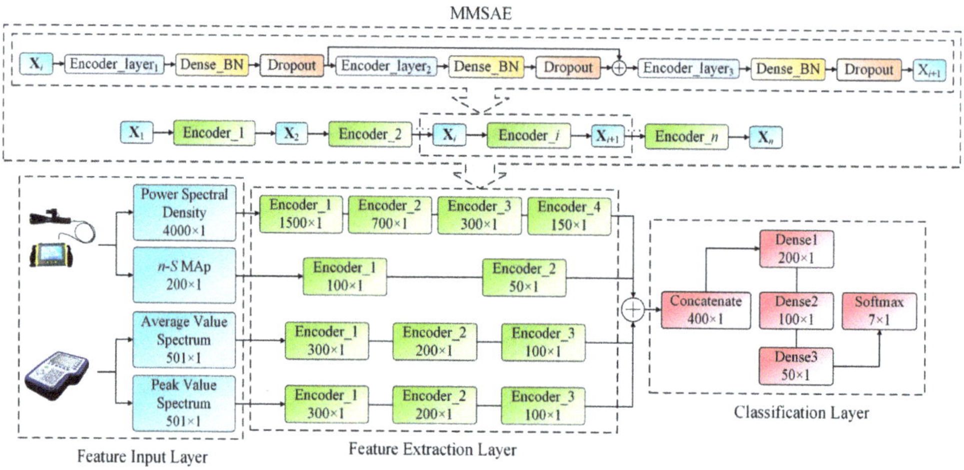

Figure 11. Overall network architecture of the proposed method.

The feature data include the electromagnetic wave peak spectrum, electromagnetic wave average spectrum, special spacing *n-S* map, and power spectral density calculated using the Blackman–Tukey method.

The feature extraction layer is constructed based on the MMSAE network. This paper first designs an auto-encoder-based multi-scale perception module and then realizes feature extraction and dimension reduction by stacking and connecting the encoding portions of multiple improved auto-encoders.

The classification layer consists of the fusion layer, the hidden layer, and the output layer. The supervised feature extraction layer is optimized, the entire network is trained using the backpropagation algorithm, and the optimal training results are stored. The entire network uses the Adam optimizer, the output layer activation function is softmax, the loss function is categorical cross-entropy, and the output is seven dimensions. The remaining layer activation functions are RELU, and the loss function is MSE.

3.2. MMSAE Network

This paper proposes a multi-scale perception module based on an auto-encoder, whose structure is shown in Figure 11. The improved auto-encoder consists of three hidden layers. After inputting the data X_i, the output of the first hidden layer Encoder_layer$_1$ and the second hidden layer Encoder_layer$_2$ are combined by the batch normalization Dense_BN layer and the Dropout layer, and then by output data X_{i+1} after passing through the third hidden layer Encoder_layer$_3$, Dense_BN layer, and Dropout layer. Among them, the batch normalization of the input of each layer can accelerate the convergence speed of the model. The Dropout layer is appropriately added to each layer, and the parameter is set to 0.3, improving the robustness and generalization ability of the network and preventing overfitting.

Finally, a complete MMSAE network is constructed by cascading the encoding parts of the aforementioned multiple improved auto-encoders, which can fully utilize the feature information of different complexities of deep and shallow layers without increasing the number of network parameters and obtaining richer multi-scale features. At the same time, it can prevent overfitting caused by too many network layers and improve classification accuracy.

3.3. *Acoustic–Electric Feature Fusion Method*

3.3.1. Electromagnetic Wave Spectrum Feature Extraction

Subtracting their respective background waves from the peak spectrum and average spectrum of all electromagnetic waves yields a matrix containing 2 × 501 × 1 corresponding data samples. Signal features are extracted using a stacked auto-encoder, and the network for feature extraction is pretrained. Each sample has two input variables and is entered separately into the SAE, each AE has four hidden layers, and each SAE contains three AEs. Thus, three decoding processes are formed. Finally, two feature extraction networks with six hidden layers are generated to reduce the input from 2 × 501 × 1 to 2 × 100 × 1.

3.3.2. Ultrasonic Feature Extraction

(1) *n-S* map.

After the signal is denoised by wavelet packet and the outliers of *S* are removed by box diagram, considering the influence of the recorded audio length on the number of envelope area *S*, the maximum and minimum values of its own data are used instead of the global maximum and minimum values when normalizing the *n* of the *n-S* map, so as to express the frequency distribution relationship of *S* with different sizes without considering the absolute number. The data dimension of the *n-S* map described previously is 200 × 1, and the data in each sample are also input into the SAE containing two AEs by the stacked auto-encoder. As a result, two decoding processes are generated, as well as an extraction network with four hidden layers. The input of 200 × 1 is reduced to 50 × 1.

(2) Power spectral density.

The power spectral density is calculated by the Blackman–Tukey method. The frequency domain is subdivided into 4000 intervals, and the input data for the stacked auto-encoder is 4000 × 1. Each SAE is composed of four AEs, resulting in four decoding processes. Finally, a network for feature extraction with eight hidden layers is generated, and the input is reduced from 4000 × 1 to 150 × 1.

3.3.3. Feature Fusion and Classification

Intact insulators and six kinds of defects are represented by 0~6. The sample number, type, and data quantity are shown in Table 3. The sample number is converted to one-hot encoding as a label. The dataset is disordered, and the training set, validation set, and test set are divided at a ratio of 6:2:2. Through the aforementioned stacked auto-encoder, the four signal features are reduced to similar sizes to prevent one signal from occupying an excessive amount of weight and weakening the effect of the other signals on the results. In addition, the feature extraction method and classification network can fuse more diagnostically relevant data and ensure the detection's comprehensiveness and precision by combining the benefits of various detection methods.

Table 3. Sample list.

Number	Type	Data Quantity
0	Intact	72
1	Conduction conductive	120
2	Conduction semi-conductive	120
3	Suspension conductive	72
4	Suspension semi-conductive	72
5	Leakage tracking	72
6	Water vapor entry	72

3.4. *Diagnostic Results and Analysis*

3.4.1. Fusion Diagnosis Results and Analysis

Figure 12 depicts the accuracy and loss-value of the algorithm during the training process. After 500 generations of training, the accuracy of the training set and the validation set converges to 1, and the loss-value curve indicates that the model training effect is satisfactory.

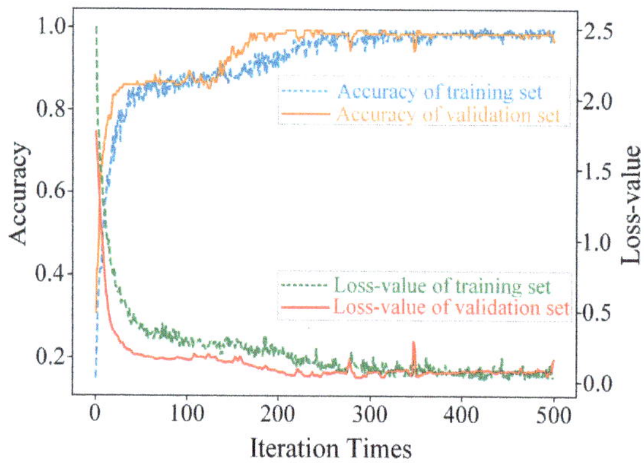

Figure 12. Accuracy and loss-value of fusion data training set and validation set.

The accuracy of the final test set obtained from the confusion matrix of Figure 13 reaches 99.17%. With strong discharge, the algorithm can accurately distinguish between conduction and suspension defects, and with weak discharge, it can distinguish between leakage tracking, water vapor entry, and an intact insulator.

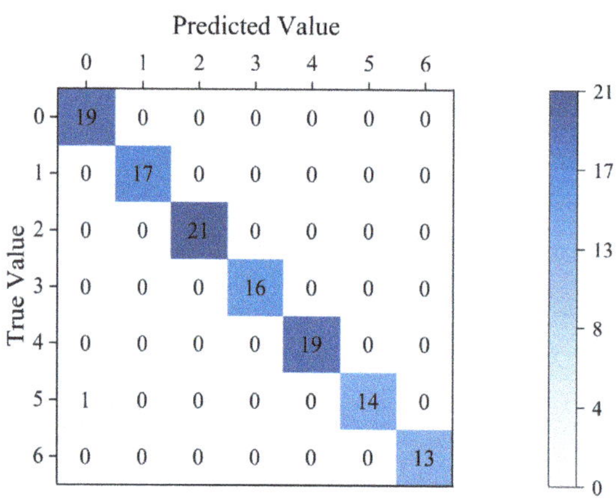

Figure 13. Confusion matrix of fusion data.

3.4.2. Comparison and Analysis of MMSAE and ANN Results

As a mainstream artificial intelligence algorithm, ANN has been widely used in the field of insulator defect classification and recognition and has a high recognition accuracy [33,34]. Therefore, under the condition of keeping the parameters of the hidden layer dimension, the number of hidden layers, the number of iterations, the optimizer, the activation function, and the objective function consistent, the ANN and the MMSAE network proposed in this paper are used to diagnose ultrasonic, electromagnetic wave spectrum and acoustic–electrical fusion, respectively. Table 4 provides the results. Among them, reasonable model evaluation metrics can quantify the classification model's performance. The main evaluation indexes are *recall*, *precision*, F_1 *score*, and *total accuracy*.

Table 4. Identification results of MMSAE and ANN.

Defect Type	Index	MMSAE			ANN		
		Fusion (%)	Ultrasonic (%)	Electromagnetic Wave (%)	Fusion (%)	Ultrasonic (%)	Electromagnetic Wave (%)
0	*Recall*	95	82.35	90	86.36	50	85
	Precision	100	73.68	94.74	100	31.58	89.47
	F_1-score	97.44	77.78	92.31	92.68	38.71	87.18
1	*Recall*	100	100	100	100	100	100
	Precision	100	94.12	100	100	100	100
	F_1-score	100	96.97	100	100	100	100
2	*Recall*	100	100	100	100	100	100
	Precision	100	100	100	100	100	100
	F_1-score	100	100	100	100	100	100
3	*Recall*	100	84.21	100	100	100	100
	Precision	100	100	100	100	100	100
	F_1-score	100	91.43	100	100	100	100
4	*Recall*	100	76.92	100	100	68.75	100
	Precision	100	52.63	100	100	57.89	100
	F_1-score	100	62.50	100	100	62.86	100
5	*Recall*	100	70.59	92.86	92.31	40.91	85.71
	Precision	93.33	80	86.67	80	60	80
	F_1-score	96.55	75	89.66	85.71	48.65	82.76
6	*Recall*	100	47.06	100	100	50	100
	Precision	100	61.54	100	92.31	61.54	100
	F_1-score	100	53.33	100	96	55.17	100
Total accuracy		99.17	80.83	97.5	96.67	73.30	95.83

The following conclusions can be obtained from the above table:

1. The training effects of the two networks are compared. As training sample sets, the fusion, ultrasonic, and electromagnetic wave data are utilized. The *total accuracy* obtained by training on the MMSAE model was 99.17%, 80.83%, and 97.5%, respectively, and that on the ANN was 96.67%, 73.30%, and 95.83%, respectively. It demonstrates that the training effect of MMSAE using any data is superior to that of ANN when using the same data.
2. Comparing the classification results of different training data, the fusion method shows the best effect on the two models, MMSAE and ANN, with 99.17% and 96.67% accuracy, respectively. The electromagnetic wave method is 97.5% and 95.83%, respectively. The ultrasonic method has the lowest effect, which is 80.83% and 73.3%, respectively. The classification results indicate that acoustic–electrical joint detection can combine the benefits of the two individual detection methods to achieve superior detection results.
3. Using the F_1 *score* to compare the recognition of various defects, MMSAE trained with fusion data has a good recognition effect on all defects. In the detection of electromagnetic wave data, type 0 and type 5 are difficult to identify. The F_1 *score* in MMSAE is 92.31% and 89.66%, respectively, and the F_1 *score* in ANN is 87.18% and 82.76%, respectively. Nonetheless, the electromagnetic wave spectrum method can achieve an overall superior classification effect. In the detection of ultrasonic data, type 0, type 4, type 5, and type 6 were difficult to identify, and the F_1 *score* in MMSAE is 77.78%, 62.50%, 75%, and 53.33%, respectively. The F_1 *score* in ANN is 38.71%, 62.86%, 48.65%, and 55.17%, respectively. This indicates that the F_1 *score* is correlated positively with the discharge intensity of each defect type.

4. Engineering Cases

This paper develops a composite insulator defect diagnosis software based on PYQT5 and Keras. The software utilizes the MMSAE network-based multi-source feature fusion diagnosis method. In order to verify the effectiveness of the proposed composite insulator defect identification method and the developed diagnostic software, ten 500 kV aging composite insulators that were replaced on-site were subjected to testing, as shown in Figure 14.

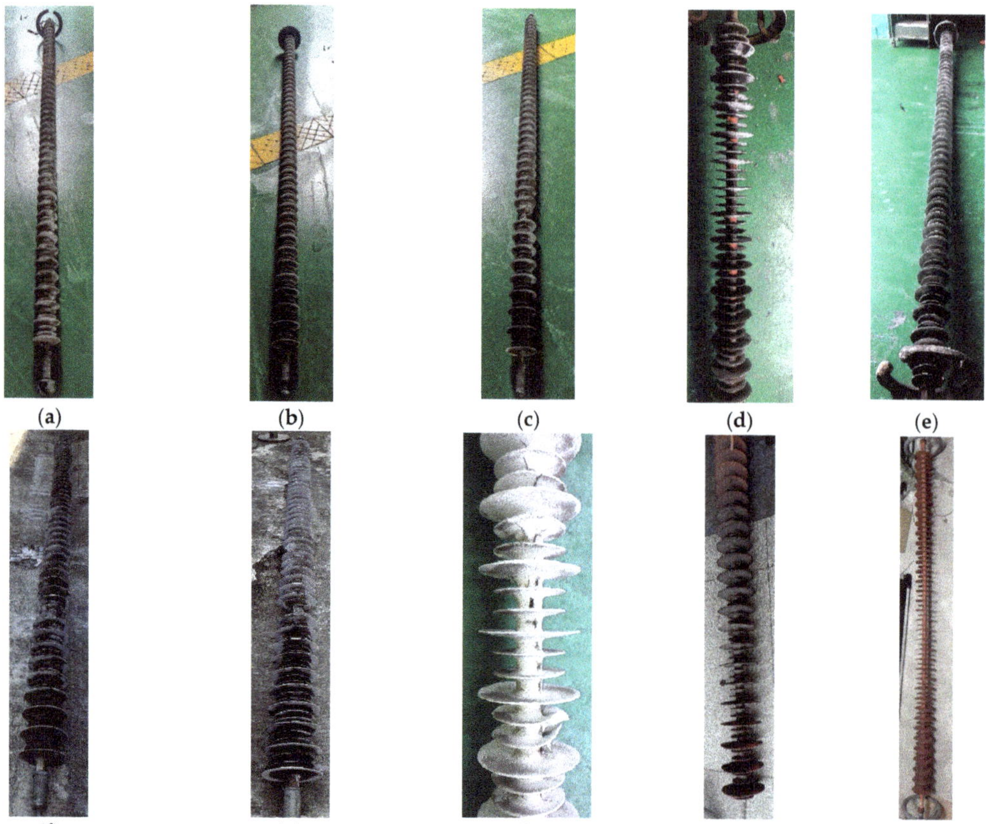

Figure 14. Field aging composite insulators: (**a**) No. 1; (**b**) No. 2; (**c**) No. 3; (**d**) No. 4; (**e**) No. 5; (**f**) No. 6; (**g**) No. 7; (**h**) No. 8; (**i**) No. 9; (**j**) No. 10.

The electromagnetic wave spectrum and ultrasonic data for the three composite insulators listed above were collected. The processed data were entered into the software for diagnosing defects in composite insulators, and the acoustic–electric fusion method was selected. The software diagnosis results are shown in Table 5.

To verify the accuracy of the software's diagnostic results, the insulators were dissected, and the core rod's surface sheath was removed. Figure 15 depicts the details of the core rod of each insulator. The dissection verification results are shown in Table 5.

Table 5. Diagnosis and dissection results of engineering cases.

No.	Software Diagnostics	Dissection Verification
1	Conduction conductive	Conduction conductive
2	Conduction semi-conductive	Conduction semi-conductive
3	Water vapor entry	Water vapor entry
4	Suspension conductive	Suspension conductive
5	Suspension semi-conductive	Suspension semi-conductive
6	Water vapor entry	Water vapor entry
7	Water vapor entry	Water vapor entry
8	Conduction conductive	Conduction conductive
9	Leakage tracking	Leakage tracking
10	Conduction semi-conductive	Conduction semi-conductive

Figure 15. Aging composite insulators after dissection: (**a**) No. 1; (**b**) No. 2; (**c**) No. 3; (**d**) No. 4; (**e**) No. 5; (**f**) No. 6; (**g**) No. 7; (**h**) No. 8; (**i**) No. 9; (**j**) No. 10.

In Figure 15, there is no adhesion between the sheath and the core rod of composite insulators No. 1, No. 2, No. 8, and No. 10, the umbrella skirt sheath is severely pulverized, and the surface of the core rod defect is yellow-white. Cutting the surface of the core rod at the defect, the glass fiber used to make the core rod immediately fractured and pulverized. The defect extends from the high-voltage end metal fittings to the low-voltage end and is located on the side of the core rod. The defects are spirally extended around the mandrel, and the extension length is No. 8 > No. 10 > No. 1 > No. 2, indicating that the longer the extension, the more serious the defects. After analysis, it is determined that composite insulators No. 1 and No. 8 have a conduction conductive defect, and No. 2 and No. 10 have a conduction semi-conductive defect.

The sheath of No. 3, No. 6, and No. 7 composite insulators closely adheres to the core rod, and the core rod defect only appears between the high-voltage end metal fittings and the first umbrella skirt. There is no obvious pulverization on the surface of the defect, which proves that the discharge is relatively weak. It has been determined that composite insulators No. 3, No. 6, and No. 7 have a water vapor entry defect.

The interface adhesion between the sheath and the core rod of composite insulators No. 4 and No. 5 is weak, the powdering of the shed sheath is more serious, and the surface of the core rod defect also shows yellow-white. The defects are located in the middle area of the mandrel and extend to both sides. The extension length of No. 4 is greater than that of No. 5, indicating that the defect is more serious. After analysis, it is determined that composite insulator No. 4 has a suspension conductive defect, and No. 5 has a suspension semi-conductive defect.

The surface of the shed sheath at the defect of composite insulator No. 9 is cracked and seriously pulverized. Under the combined action of electric field and electrolyte, the surface of the core rod is partially carbonized, and finally, a conductive path is formed. After analysis, it was determined that composite insulator No. 9 has a leakage tracking defect. The consistency between software diagnosis and anatomical analysis results validates the effectiveness of the MMSAE algorithm and defect-recognition software.

5. Conclusions

In view of the limitations of existing live defect-detection methods for composite insulators, this paper proposes a multi-scale perception multi-input deep learning network and develops a composite insulator defect diagnosis model based on acoustic–electric joint detection. The following are the conclusions:

1. Based on the discharge characteristics of defective composite insulators, six common defects are simulated artificially. The electromagnetic wave peak spectrum, electromagnetic wave average spectrum, ultrasonic *n-S* map, and power spectral density of various defects can distinguish composite insulators with various defects.
2. The network architecture of MMSAE is proposed based on acoustic and electrical information. First, the multi-source input data are pre-trained, then the output is fused, and finally, the final fine-tuning is performed to solve the problem that different types of field-based detection methods cannot share information. At the same time, the feature extraction method and classification network are extensible, allowing for the fusion of additional diagnostically useful data. By combining the benefits of various detection methods, the comprehensiveness and accuracy of detection are ensured.
3. Based on the model evaluation index's *total accuracy* and F_1 *score*, it is demonstrated that acoustic–electric fusion is more effective than the use of two detection methods separately. In the ultrasonic method, electromagnetic wave spectrum method, and acoustic–electric fusion method, it is demonstrated that MMSAE achieves greater accuracy than conventional ANN networks.
4. The composite insulator defect diagnosis software is used to identify the defects of the aging composite insulator on site. The consistency between the software diagnosis results and the anatomical analysis results validates the efficacy and applicability of the MMSAE algorithm and the defect identification software.

The method proposed in this paper can fully serve the state detection of transmission lines. In future research, improvement and in-depth mining analysis of the technology can be carried out from the following two aspects: on the one hand, with the emergence of more detection methods, the algorithm structure and label categories can be expanded. On the other hand, by adding more environmental factors to the training samples, the feature set can be continuously expanded to improve the sample library. Finally, it provides ideas for the multi-source feature fusion composite insulator online diagnosis system with more perfect functions, more accurate identification and diversification.

Author Contributions: Conceptualization, X.W.; methodology, J.X.; software, C.C.; validation, X.W., J.X. and C.F.; formal analysis, C.C. and S.S.; investigation, B.Z.; writing—original draft preparation, B.Z.; writing—review and editing, S.S.; visualization, J.X.; supervision, C.F. All authors have read and agreed to the published version of the manuscript.

Funding: This work was supported by the National Natural Science Foundation of China (no. 52207150).

Data Availability Statement: Not applicable.

Conflicts of Interest: The authors declare no conflict of interest.

References

1. Liang, X.; Gao, Y.; Wang, J.; Li, S. Rapid development of silicone rubber composite insulator in China. *High Volt. Eng.* **2016**, *42*, 2888–2896. [CrossRef]
2. Chen, C.; Shu, S.; Dong, Y.; Wang, J.; Jin, M. Partial discharge pattern classification of composite insulators by electromagnetic spectrum and stacked autoencoder network. In Proceedings of the 22nd International Symposium on High Voltage Engineering (ISH), Xi'an, China, 21–25 November 2021; pp. 1124–1129. [CrossRef]
3. Xia, Y.; Song, X.; He, J.; Jia, Z.; Wang, X. Evaluation method of aging for silicone rubber of composite insulator. *Trans. China Electrotech. Soc.* **2019**, *34*, 440–448. [CrossRef]
4. Fu, P.; Zhao, Z.; Li, X.; Cui, X.; Wen, T.; Yang, Z.; Mo, S.; Zhang, P. Partial discharge measurement and analysis in PPIs. *IET Power Electron.* **2019**, *12*, 138–146. [CrossRef]
5. Phillips, A.; Kuffel, J.; Baker, A.; Burnham, J.; Carreira, A.; Cherney, E. Electric fields on AC composite transmission line insulators. *IEEE Trans. Power Delivery* **2008**, *23*, 823–830. [CrossRef]
6. Amin, M.; Amin, S.; Ali, M. Monitoring of leakage current for composite insulators and electrical devices. *Rev. Adv. Mater. Sci.* **2009**, *21*, 75–89.
7. Tian, F.; Hao, Y.; Zou, Z.; Zheng, Y.; He, W.; Yang, L.; Li, L. An ultrasonic pulse-echo method to detect internal defects in epoxy composite insulation. *Energies* **2019**, *12*, 4804. [CrossRef]
8. Li, S.; Li, J. Condition monitoring and diagnosis of power equipment: Review and prospective. *High Volt.* **2017**, *2*, 82–91. [CrossRef]
9. Raymond, W.; Illias, H.; Bakar, A.; Mokhlis, H. Partial discharge classifications: Review of recent progress. *Measurement* **2015**, *68*, 164–181. [CrossRef]
10. Li, J.; Han, X.; Liu, Z.; Li, Y. Review on partial discharge measurement technology of electrical equipment. *High Volt. Eng.* **2015**, *41*, 2583–2601. [CrossRef]
11. Yaacob, M.; Alsaedi, M.; Rashed, J.; Dakhil, A.; Atyah, S. Review on partial discharge detection techniques related to high voltage power equipment using different sensors. *Photonic Sens.* **2014**, *4*, 325–337. [CrossRef]
12. Li, X.; Meng, F.; Ji, S.; Wang, S.; Wang, F.; Li, Y. On-line detecting system for composite insulator based on pulse current method. *High Volt. Eng.* **2006**, *32*, 44–47. [CrossRef]
13. Uckol, H.; Karaca, B.; Ilhan, S. DC and AC electric field analysis and experimental verification of a silicone rubber insulator. *Electr. Eng.* **2020**, *102*, 503–514. [CrossRef]
14. Li, C.; Niu, B.; Zeng, R.; Yu, Z.; Zhang, F. Detection of internal insulation defects of composite insulator based on electro-optic electric field sensor. *High Volt. Eng.* **2014**, *40*, 2422–2428. [CrossRef]
15. Li, J.; Jiang, T.; Harrison, R.; Grzybowski, S. Recognition of ultra high frequency partial discharge signals using multi-scale features. *IEEE Trans. Dielectr. Electr. Insul.* **2012**, *19*, 1412–1420. [CrossRef]
16. Yuan, C.; Xie, C.; Li, L.; Zhang, F.; Gubanski, S. Ultrasonic phased array detection of internal defects in composite insulators. *IEEE Trans. Dielectr. Electr. Insul.* **2016**, *23*, 525–531. [CrossRef]
17. Gao, K.; Lyu, L.; Huang, H.; Fu, C.; Chen, F.; Jin, L. Insulation defect detection of electrical equipment based on infrared and ultraviolet photoelectric sensing technology. In Proceedings of the IECON 2019-45th Annual Conference of the IEEE Industrial Electronics Society, Lisbon, Portugal, 14–17 October 2019; pp. 2184–2189. [CrossRef]
18. Jain, A.; Duin, R.; Mao, J. Statistical pattern recognition: A review. *IEEE Trans. Pattern Anal. Mach. Intell.* **2002**, *22*, 4–37. [CrossRef]
19. Abramson, N.; Braverman; Sebestyen, G. *Pattern Recognition and Machine Learning*; IEEE: Piscataway, NJ, USA, 1963; Volume 9, pp. 257–261.
20. Orriols-Puig, A.; Casillas, J.; Bernadó-Mansilla, E. Genetic-based machine learning systems are competitive for pattern recognition. *Evol. Intell.* **2008**, *1*, 209–232. [CrossRef]
21. Bulgarevich, D.; Tsukamoto, S.; Kasuya, T.; Demura, M.; Watanabe, M. Pattern recognition with machine learning on optical microscopy images of typical metallurgical microstructures. *Sci. Rep.* **2018**, *8*, 2078. [CrossRef] [PubMed]
22. Jing, X.; Peng, X.; Jiang, W.; Zhou, W.; Zhou, C.; Tang, Z. Phase resolved partial discharge pattern recognition method for online cable condition monitoring system based on K-Means clustering. *High Volt. Eng.* **2012**, *38*, 2437–2446.

23. Hunter, J.; Hao, L.; Lewin, P.; Evagorou, D.; Kyprianou, A.; Georghiou, G. Comparison of two partial discharge classificatio methods. In Proceedings of the 2010 IEEE International Symposium on Electrical Insulation, San Diego, CA, USA, 6–9 June 201 pp. 1–5. [CrossRef]
24. Iorkyase, E.; Tachtatzis, C.; Glover, I.; Lazaridis, P.; Upton, D.; Saeed, B.; Atkinson, R. Improving RF-based partial discharg localization via machine learning ensemble method. *IEEE Trans. Power Delivery* **2019**, *34*, 1478–1489. [CrossRef]
25. Abdel-Galil, T.; Sharkawy, R.; Salama, M.; Bartnikas, R. Partial discharge pattern classification using the fuzzy decision tre approach. *IEEE Trans. Instrum. Meas.* **2005**, *54*, 2258–2263. [CrossRef]
26. Gao, A.; Zhu, Y.; Cai, W.; Zhang, Y. Pattern recognition of partial discharge based on VMD-CWD spectrum and optimized CN with cross-layer feature fusion. *IEEE Access* **2020**, *8*, 151296–151306. [CrossRef]
27. Gao, J.; Zhu, Y.; Zheng, Y.; Jia, Y. Pattern recognition of partial discharge based on VMD-WVD and SSAE. *Proc. CSEE* **2019**, *3* 4118–4129. [CrossRef]
28. Contin, A.; Cavallini, A.; Montanari, G.; Pasini, G.; Puletti, F. Digital detection and fuzzy classification of partial discharge signal *IEEE Trans. Dielectr. Electr. Insul.* **2002**, *9*, 335–348. [CrossRef]
29. Ardila-Rey, J.; Ortiz, J.; Creixell, W.; Muhammad-Sukki, F.; Bani, N. Artificial generation of partial discharge sources through a algorithm based on deep convolutional generative adversarial networks. *IEEE Access* **2020**, *8*, 24561–24575. [CrossRef]
30. Liu, L.; Wang, L.; Mei, H.; Guo, C. Defects detection methods for composite insulators based on shearing speckle interferometr *Proc. CSEE* **2019**, *39*, 4599–4606. [CrossRef]
31. Ramesh, M.; Cui, L.; Gorur, R. Impact of superficial and internal defects on electric field of composite insulators. *Int. J. Elect Power Energy Syst.* **2019**, *106*, 327–334. [CrossRef]
32. Lv, F.; Liu, J.; Shi, P.; Wang, F.; Xiong, J.; Li, C. Experimental research on composite insulator fault detection. *High Volt. Appa* **2013**, *49*, 95–100. [CrossRef]
33. Polisetty, S.; El-Hag, A.; Jayram, S. Classification of common discharges in outdoor insulation using acoustic signals and artificia neural network. *High Volt.* **2020**, *4*, 333–338. [CrossRef]
34. Ferrah, I.; Chaou, A.K.; Maadjoudj, D.; Teguar, M. Novel colour image encoding system combined with ANN for discharge pattern recognition on polluted insulator model. *IET Sci. Meas. Technol.* **2020**, *14*, 718–725. [CrossRef]

energies

MDPI

Article

Fault Diagnosis for Motor Bearings via an Intelligent Strategy Combined with Signal Reconstruction and Deep Learning

Weiguo Li [1], Naiyuan Fan [2], Xiang Peng [1], Changhong Zhang [1], Mingyang Li [1], Xu Yang [2] and Lijuan Ma [2,*]

[1] Ultra High Voltage Transmission Company Electric Power Research Institute, China Southern Power Grid Company Limited, Guangzhou 510000, China; liweiguo2005@163.com (W.L.); 13503019667@126.com (X.P.); longrainbow@163.com (C.Z.); limyhn@126.com (M.L.)

[2] Henan Pinggao Electric Co., Ltd., Pingdingshan 476000, China; fannaiyuan@foxmail.com (N.F.); yangxuww@outlook.com (X.Y.)

* Correspondence: 13071812107@163.com

Abstract: To overcome the incomplete decomposition of vibration signals in traditional motor-bearing fault diagnosis algorithms and improve the ability to characterize fault characteristics and anti-interference, a diagnostic strategy combining dual signal reconstruction and deep learning architecture is proposed. In this study, an improved complete ensemble empirical mode decomposition with adaptive noise (CEEMDAN) and variational mode decomposition (VMD)-based signal reconstruction method is first introduced to extract features representing motor bearing faults. A feature matrix construction method based on improved information entropy is then proposed to quantify these fault features. Finally, a fault diagnosis algorithm architecture integrating a multi-scale convolutional neural network (MSCNN) with attention mechanisms and a bidirectional long short-term memory network (BiLSTM) is developed. The experimental results for four fault states show that this model can effectively extract fault features from original vibration signals and, compared to other fault diagnosis models, offer high diagnostic accuracy and strong generalization, maintaining high accuracy even under varying speeds and noise interference.

Keywords: motor bearings; fault diagnosis; feature extraction; signal reconstruction; deep learning

1. Introduction

Electric motors are ubiquitous in the realm of rotating machinery and find extensive application across electrical and energy-related industries. However, their operation often occurs under harsh conditions, making them prone to failures that can lead to shutdowns and pose significant threats to production safety. To ensure safe and continuous operations while mitigating economic losses associated with maintenance and repair, rapid and accurate fault diagnosis of motor bearings is essential [1,2].

Traditionally, fault diagnosis methods for electric motor bearings have relied on signal processing techniques [3–5]. These methods typically analyze vibration signals or current signals in the time domain, frequency domain, or both, to diagnose bearing faults. Commonly used signal processing methods include empirical mode decomposition (EMD) [6], variational mode decomposition (VMD) [7], and wavelet transform [8]. While EMD is capable of processing non-linear and non-stationary signals using adaptive basis functions, it may suffer from mode mixing, which can degrade the accuracy of signal decomposition [9]. Ensemble empirical mode decomposition (EEMD) addresses this issue by effectively suppressing mode mixing [10]. VMD, as a newer adaptive signal processing method, transforms the signal decomposition process into a variational problem, thereby achieving adaptive separation of signals without mode mixing or endpoint effects [11,12]. Despite these advancements, the non-stationary and non-linear characteristics of real-world bearing vibration signals present challenges for feature extraction. As a result, traditional signal processing approaches alone make it hard to provide adaptive fault diagnosis. The development of fault classification methods integrating multiple methodologies has become a leading direction in this field.

With the advent of artificial intelligence, deep neural networks have demonstrated unique advantages in handling data with non-linear relationships. Numerous studies have applied deep learning algorithms to establish mappings between signals and faults, facilitating fault diagnosis for electric motor bearings [13]. The limitations of traditional CNNs have led to the emergence of a new type of CNN called CapsNet, which overcomes the problems of CNN position information loss and feature reuse [14]. Additionally, Ref. [15] indicated that traditional deep learning-based bearing fault diagnosis methods assume that the training and testing data follow the same distribution. However, in practical scenarios, the distribution of bearing data may vary, invalidating this assumption and leading to a significant decline in diagnostic performance. To address this issue, the concept of transfer learning has been introduced in deep learning, allowing knowledge gained from other data or models to be transferred to the current task, thereby effectively resolving this problem. Similarly, algorithm architectures based on combined recurrent networks have also been proposed to overcome similar problems. For instance, a combined Convolutional Neural Network (CNN) and Long Short-Term Memory (LSTM) model was proposed for bearing fault diagnosis [16] using time–frequency images and vibration signals as inputs for multi-dimensional feature extraction and fusion. Another study introduced a hybrid CNN-LSTM neural network for end-to-end rolling bearing fault diagnosis [17]. Yet another approach combined the CNN and bidirectional LSTM (BiLSTM) to adaptively extract features for bearing fault diagnosis [18]. Although these integrated methods have shown promise, they still face challenges in complex vibration environments, including incomplete signal decomposition, weak correlation between defect features and signals, and limitations in model interpretability and generalization.

To address these issues, we propose a novel fault diagnosis model based on signal reconstruction and feature dependency. In our research, we first combine Improved Complete Ensemble Empirical Mode Decomposition with Adaptive Noise (ICEEMDAN) and VMD to extract effective vibration signal components under complex noise conditions. We then introduce an improved signal entropy-based method to quantify defect features, yielding parameters corresponding to different fault states. Finally, we utilize attention mechanisms to enhance the interpretability and generalization capabilities of the MSCNN-MSBiLSTM model for fault diagnosis. Our results demonstrate that the proposed model can adaptively adjust its parameters based on signal features, achieving a diagnostic accuracy of 95% under laboratory conditions. Thus, the methodology presented here offers a new algorithmic framework for the fault diagnosis of electric motor bearings.

2. Experimental Platform and Data Description

To substantiate the efficacy and practical applicability of the proposed diagnostic methodology, this study analyzes a publicly available dataset provided by Case Western Reserve University (CWRU). The CWRU rolling element bearing fault simulation test rig, depicted in Figure 1a, was utilized [19].

Data acquisition was performed at a sampling frequency of 15 kHz, with the motor operating at a speed of 500, 700, and 900 rpm. Three distinct fault diameters of 0.21 mm, 0.33 mm, and 0.52 mm were simulated, representing inner race, outer race, and rolling element faults, respectively, along with a healthy condition. A sling window technique with a fixed sample length (4096) and a fixed sampling interval (2000) was employed to extract signal samples from the CWRU dataset.

Specifically, the vibration signals generated by the motor's rolling bearings constitute a one-dimensional time series with temporal characteristics. The raw measurement signals corresponding to the four conditions, including healthy (HEA), inner race fault (IRF), outer race fault (ORF), and rolling element fault (REF) are illustrated in Figure 1b. The data used in the current database are collected under the same platform and conditions, so the samples in the dataset follow the same generation process or distribution. To evaluate its homogeneity and representativeness, the mean standard deviation is performed on the signals of 4000 samples (including four types of states), and the results are shown in Figure 1c.

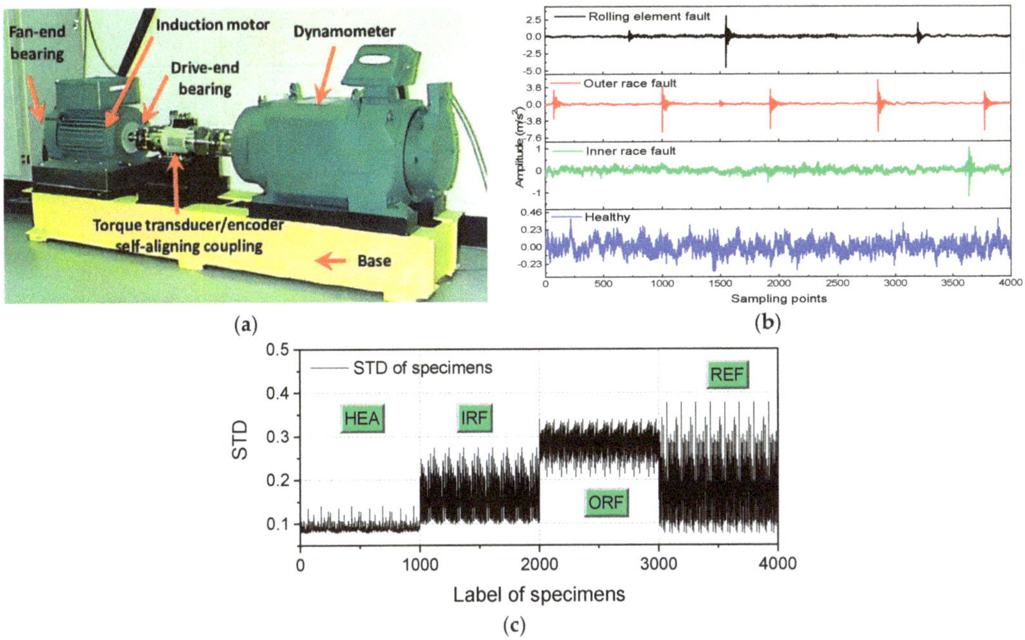

Figure 1. (**a**) CWRU rolling bearing for fault simulation test bench in [17], (**b**) original signal of rolling bearing corresponding different states, (**c**) the standard deviation of all specimens.

Additionally, the original data are divided into datasets Φ_1, Φ_2, and Φ_3 according to the three speeds. Then, for each given speed, 1000 samples were randomly selected and subsequently divided into training, testing, and verifying sets at a 7:2:1 ratio. This process then yielded a training set consisting of 2800 samples and a testing/verifying set containing 800/400 samples, both of which included all four defect categories (HEA, IRF, ORF, and REF), as tabulated in Table 1.

Table 1. Details of the training and test corresponding to various speeds.

Dataset	Training Set	Test Set	Verification Set	State
$\Phi_1/\Phi_2/\Phi_3/\Phi_4$	175	50	25	HEA
	175	50	25	IRF
	175	50	25	ORF
	175	50	25	REF
Total	2800	800	400	

3. Signal Decomposition and Feature Extraction Model

3.1. Signal Reconstruction Method Based on ICEEMDAN and VMD

Step I. Acquisition of high-frequency characteristics of signals

CEEMDAN (Complete Ensemble Empirical Mode Decomposition with Adaptive Noise) is a signal processing method designed to decompose complex signals into several Intrinsic Mode Functions (IMFs) [20]. While CEEMDAN reduces modal aliasing, it faces challenges in the early stages of decomposition due to the presence of significant noise and similar-scale features in the fault signals, which hinder the precise selection of IMFs.

Kernel Principal Component Analysis (KPCA) can effectively extract non-linear features and reduce dimensionality by mapping data into a high-dimensional feature space, thereby efficiently handling complex datasets [21].

Therefore, incorporating KPCA into the CEEMDAN framework not only preserves the non-linear characteristics of the data but also eliminates redundant information, aiding in the extraction of more representative and discriminative non-linear principal components. This paper employs cross-analysis to determine the optimal component set, ensuring the retention of critical information during dimensionality reduction and the extraction of appropriate fractal dimensions of non-linear principal components. The flowchart of the improved CEEMDAN algorithm is presented in Figure 2.

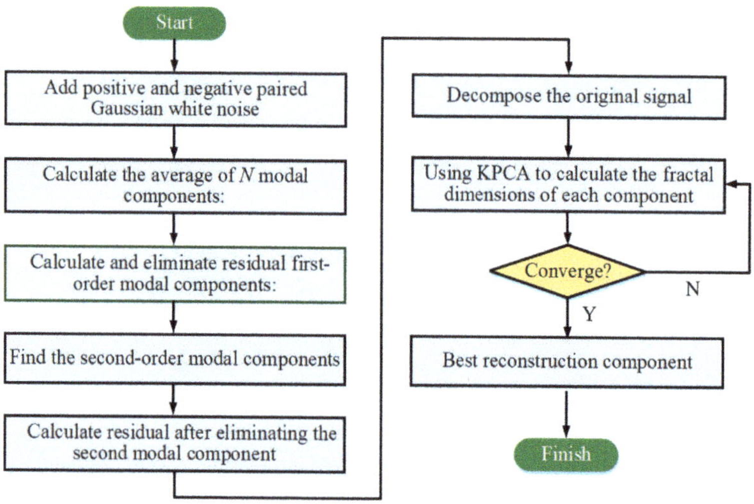

Figure 2. Computation scheme of ICEEMDAN.

For bearing vibration signals, the first IMF component derived from the ICEEMDAN algorithm primarily encapsulates the high-frequency elements or the most dynamic portions of the original signal. Since high-frequency components are indicative of mechanical faults, the first IMF is crucial for early fault detection and diagnosis. The outcomes of signal decomposition for the four states using ICEEMDAN are illustrated in Figure 3 (one sample from each state listed in Table 1 was randomly chosen for demonstration, and IMFs are preset as 6).

Step II. Acquisition of high-frequency characteristics of signals

Since the first IMF from ICEEMDAN encompasses high-frequency components, VMD can further separate distinct characteristic frequencies within this high-frequency information [7], offering a more precise spectral analysis. This additional decomposition aids in clearly identifying these features, thereby improving the accuracy of fault detection. Additionally, the VMD algorithm effectively distinguishes between different frequency components, aiding in noise reduction and preserving more relevant signal features. The detailed V-IMF components derived from the VMD decomposition of IMF1, as illustrated in Figure 3, are shown in Figure 4 (the number of V-IMF components is preset as 7).

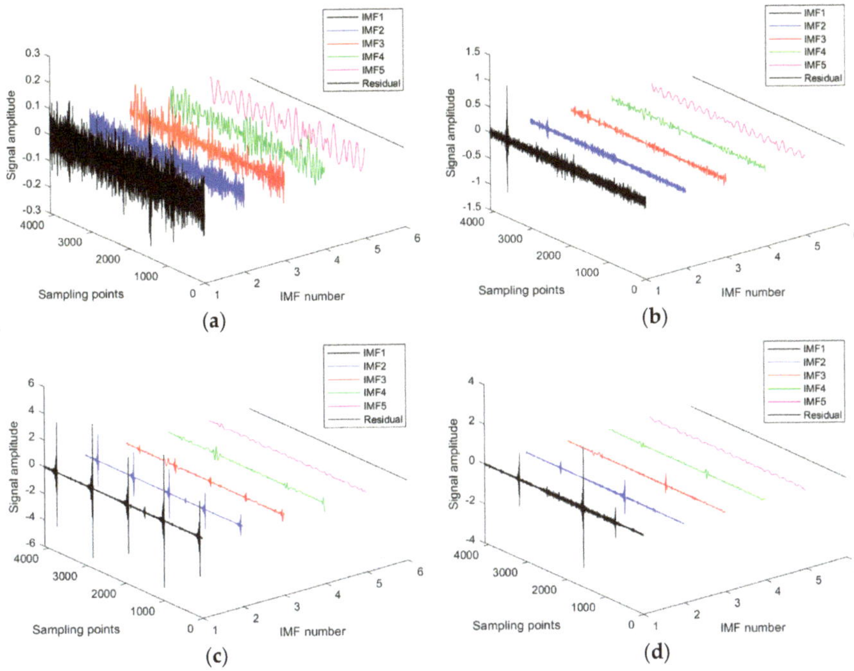

Figure 3. Signal decomposition results via ICEEMDAN. (**a**) HEA, (**b**) IRF, (**c**) ORF, and (**d**) REF.

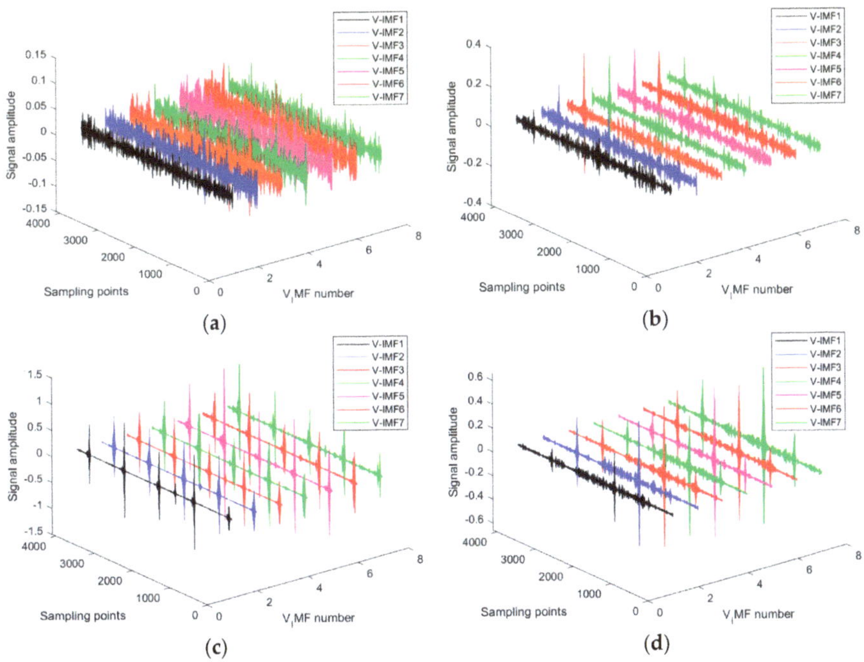

Figure 4. Sub−signal decomposition results (V−IMF) via VMD. (**a**) HEA, (**b**) IRF, (**c**) ORF, and (**d**) REF.

3.2. Fault Feature Extraction Method Based on Improved Information Entropy (IIE)

Fault signals often introduce non-periodic or abrupt components into the signal. Information entropy highlights these anomalies, assesses the significance of various modes, and determines the purity of the signal, thereby assisting in fault diagnosis [22]. For the IMF components of vibration signals, a more ordered system corresponds to lower entropy values. The specific method for calculating information entropy is as follows.

For any variable X, the expression for information entropy $H(X)$ is as follows:

$$H(X) = -\sum_{i=1}^{n} p_i \lg p_i \tag{1}$$

In (1), p_i is the ratio of the energy of the i-th IMF component to the total energy, and the IMF component is obtained from VMD (in Figure 4). The calculation formula for p_i is as follows:

$$\begin{aligned} & p_i = \tfrac{E_i}{E}, E = \sum_{i=1}^{n} E_i \\ & s.t.\ E_i = \int_{-\infty}^{\infty} x^2(t)\mathrm{d}t = \int_{-\infty}^{\infty} \left[\sum_{i=1}^{n} x_i(t)\right]^2 \mathrm{d}t \end{aligned} \tag{2}$$

where $x(t)$ is the input signal sequence concerning time t and X_i is the i-th signal.

When information entropy is used to describe bearing motion, fault-induced periodic impacts lead to a more ordered signal. Thus, IMF components containing fault signals will exhibit lower entropy values. However, during the early stages of bearing faults, the features may be less pronounced, resulting in lower identification rates. To address this, improvements to entropy are necessary, such as amplifying individual IMF features.

Additionally, kurtosis, an important indicator for describing waveform peaks and evaluating fault impacts, can enhance the model's sensitivity to impact information. The entropy can be refined using kurtosis and is defined as follows:

$$Q_i = \frac{K_i}{H_i} \tag{3}$$

In (3), Q_i is the improved information entropy of the i-th IMF component and H_i and K_i are the information entropy and kurtosis indicators of the i-th IMF component, respectively. Since information entropy only represents the corresponding amount of IMF information, and due to environmental noise and other reasons, fault information does not show obvious entropy values.

Then, the number of IMFs decomposed from each specific signal segment is constant, and the improved information entropy of IMFs will have an upper limit value of the following:

$$\max(Q) = -\ln\frac{1}{n} \tag{4}$$

In (4), n is the number of IMFs obtained by decomposition and Q is information entropy. The average improved information entropy of a single set of original signals is as follows:

$$\bar{Q} = \frac{\iint\cdots\int Q\,\mathrm{d}Q_1\,\mathrm{d}Q_2\cdots\mathrm{d}Q_n}{\iint\cdots\int \mathrm{d}Q_1\,\mathrm{d}Q_2\cdots\mathrm{d}Q_n} = \sum_{i=2}^{n}\frac{Q_i}{2n} \tag{5}$$

Compared to the unmodified entropy in (3) to (5), IIE uses kurtosis to improve the model's sensitivity to impact information and address the issue of low recognition rates of entropy in less distinct fault components. Finally, relying on the above calculation process, the information entropy corresponding to the V-IMF obtained from VMD decomposition is calculated and used as the feature input (4000 × 7) for training CNN BiLSTM with different samples.

To examine the discrimination of these features in the case of state evaluation, t-distributed stochastic neighbor embedding (t-SNE) [23] is thus used for visualizing the formed feature input, and the resulting pictures are presented in Figure 5. The t-SNE algorithm, known for its efficiency and visualization capabilities in dimensionality reduction, aims to visualize the discrimination of features corresponding to different fault states. In Figure 5, 4000 signals with improved entropy (fault features, seven dimensions) are used as input, resulting in a three-dimensional fused feature representation (which lacks specific physical meaning). The visualization in Figure 5 shows that the signal features of different fault states (indicated by different colors) exhibit distinct regional distributions.

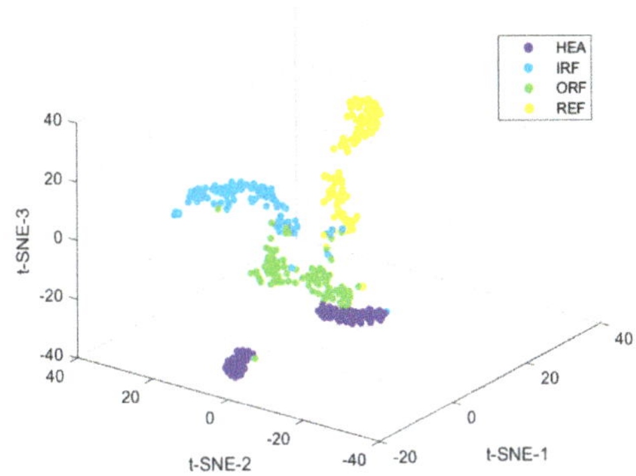

Figure 5. The t-SNE results for visualizing the discrimination of feature input corresponding to various states.

4. Fault Diagnosis Model Based on Data-Driven and Deep Learning

4.1. MSCNN-BiLSTM-Based Fault Diagnosis Model Integrating an Attention Mechanism

In this study, spatial features are initially extracted using a feature selection layer and a multi-scale feature extraction layer, followed by temporal feature extraction through a BiLSTM layer. A self-attention mechanism is then introduced for adaptive fusion of all features, culminating in fault diagnosis results via a Softmax layer. The preliminary feature screening significantly impacts model performance. Bearing fault features are typically low-frequency signals; thus, larger convolution kernels are employed to enhance low-frequency feature extraction and suppress high-frequency noise, while also accommodating the periodic nature of vibration signals by providing a broader receptive field.

Consequently, the model utilizes large convolution kernels for initial feature screening, followed by batch normalization and max pooling layers to accelerate convergence and mitigate gradient issues. An optimized Inception module is employed to construct a multi-scale feature extraction module, thereby improving feature extraction capabilities at various scales. To further refine the network structure and enhance fault diagnosis accuracy, batch normalization is applied after each convolution operation to expedite training and prevent overfitting. Additionally, channel attention modules are integrated with post-post-feature extraction to generate and continuously optimize channel-specific weights. The self-attention mechanism effectively mitigates redundant feature interference and prevents loss of critical information, thus improving diagnostic performance. The mathematical expression for this process is as follows:

$$A_{\mathrm{att}}(Q,P,V) = \mathrm{Softmax}\left(\frac{f(Q,P^{\mathrm{T}})}{\sqrt{d_{\mathrm{p}}}}\right)V \tag{6}$$

where Q, P, and V represent the query value, calculated value, and weight, respectively; $f(Q, P^T)$ is used to calculate the similarity between Q and P; d_p is the feature vector dimension; and Softmax is a normalization function.

The feature extraction subnetwork and channel attention modules are alternately arranged to construct a multi-scale feature extraction module. This module continuously refines fault data by emphasizing critical features and suppressing noise, thereby enhancing diagnostic efficiency. Subsequently, features from different channels are merged and processed through a max pooling layer to retain essential characteristics and reduce model parameters. Spatial features extracted from the multi-scale feature extraction module are then fed directly into the BiLSTM layer to capture temporal information from vibration signals. Outputs from the BiLSTM layer are input to a self-attention mechanism to improve the network's focus on global key features. Dimensionality reduction and information integration are performed via global average pooling, and the fault diagnosis results are ultimately produced by a Softmax classifier.

4.2. Computation Scheme and Parameter Values

Based on the previous sections, high-frequency signal components used to characterize fault features were obtained through ICCEMDAN and VMD decomposition. Subsequently, improved entropy was employed to achieve quantitative descriptions of different fault signals. Finally, a fault state evaluator was established using the entropy feature matrix based on the algorithm framework shown in Figure 6, with the entire algorithm framework illustrated in Figure 7.

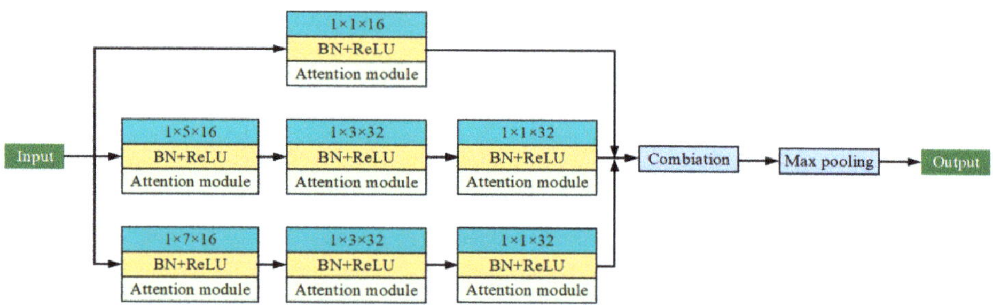

Figure 6. Multi-scale feature extraction module structure.

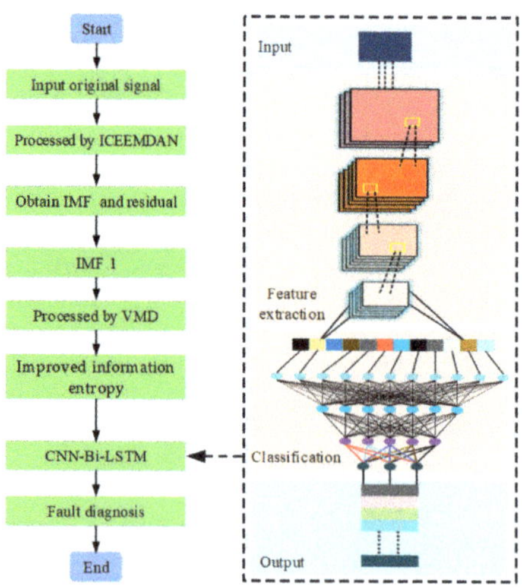

Figure 7. Entire algorithm framework for fault diagnosis.

In the current work, the main hyper-parameter (HP) includes the number of IMFs in both ICEEMDAN and VMD, as well as the learning rate of MSCNN-BiLSTM. It can significantly affect the overall performance of algorithms; the selection of the optimal value of them is thus rather fundamental. Given the small parameter range and only three optimization variables in this case, iterative computation to determine optimal parameters is more suitable.

Therefore, we use ICEEMDAN and VMD to decompose all signals (4000 signals, the length of the signal is 4096), with the mean residuals serving as criteria for selecting the number of decompositions; accuracy on the validation set (400 signals) is used to choose the learning rate for MSCNN-BiLSTM. The criteria values corresponding to different parameters are shown in Figure 8. Under the condition of ensuring the highest accuracy of signal decomposition and the evaluation results, the number of IMFs in ICEEMDAN and VMD is five and seven, respectively, and the learning rate for MSCNN-BiLSTM is 0.006.

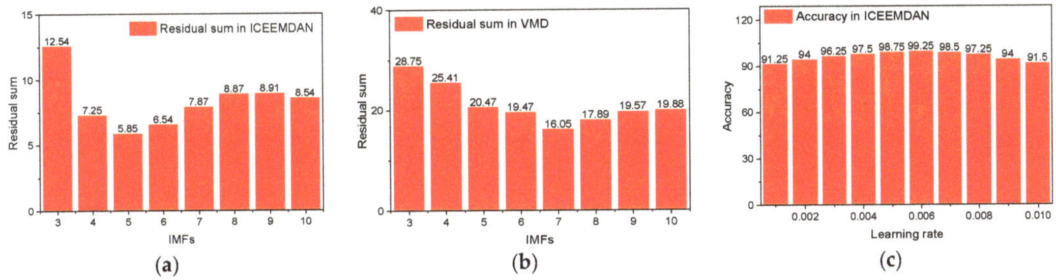

Figure 8. The criteria values correspond to different parameters: (**a**,**b**) are the number of IMFs, (**c**) is the learning rate.

Given that model parameters significantly impact signal processing results and the final evaluation outcomes, the parameters listed in Table 2 were used throughout all sample data preprocessing, feature extraction, and training. Finally, Figure 9 presents the accuracy (using a test set in Table 1) during the 50 iterations, in which the computation is completed by a computer with a 2nd Gen Intel(R) Core(TM) i9-12900K CPU at 3.20 GHz and a 64 bit Win 11 system with 16.0 GB of RAM. The Matlab Deep Learning Toolbox is used in this work for fault diagnosis. The computation time is 144 s. Obviously, the final accuracy of the test set is close to 98%.

Table 2. Parameter values used in the contained models.

Parameters used in ICCEMDAN			
Number of IMFs 5	STD of added noise 2%	Ensemble members 30	Window length 1/4
Parameters used in VMD			
Number of IMFs 7	Penalty factor 10	Convergence threshold 1×10^{-6}	Iterations 50
Parameters used in MSCNN-BiLSTM			
Minimum batch size 64	Iterations 50	Learning rate 0.006	Solver "adam"
Number of states 4	Gradient threshold 1	Learning rate decline cycle 20	Learning rate decline factor 0.25

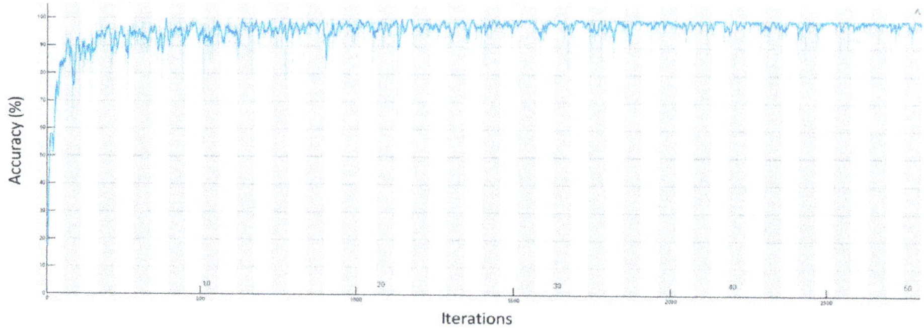

Figure 9. The diagnosis accuracy during the iterations.

5. Verification for Fault Diagnosis

5.1. Diagnosis Results Analysis

The training and testing sets with sample sizes of 2800 and 800, respectively, are used to train the fault assessment for motor bearings in four different states. The internal parameters and training process of the evaluator are detailed in Table 2 and Figure 9. Additionally, a validation set with a sample size of 400 (details in Table 3) is used to assess the accuracy of the constructed evaluator. First, the ICCEMDAN decomposition is applied to the 400 validation vibration signals based on the computation process shown in Figure 8. Then, the first IMF component obtained is further decomposed using VMD to obtain 7 V-IMF components. Next, the improved entropy of each V-IMF is calculated and used as feature inputs for the fault evaluator. Finally, the evaluation state is obtained through the CNN-BiLSTM model with an attention mechanism. Figure 10a shows the final confusion matrix, and Figure 10b compares the evaluations and actual states for different fault conditions. It can be seen that the reported evaluator achieves an overall assessment accuracy of 99.25%, with fault assessment accuracy reaching 100% for HEA and ORF states.

Table 3. Details of the used verification set.

	HEA	IRF	ORF	REF
Number	89	104	101	106
Labeled as	1	2	3	4

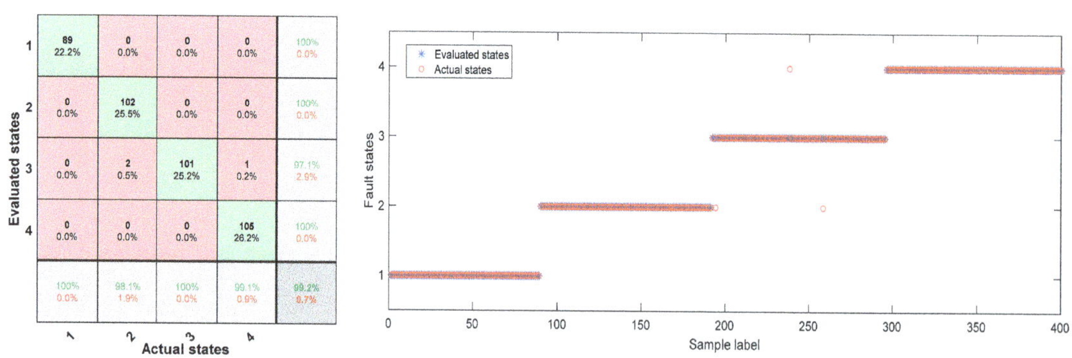

(a) (b)

Figure 10. (**a**) Confusion matrix of the evaluation results. (**b**) Comparison of the evaluated and actual states.

5.2. Comparison of Fault Diagnosis Accuracy via Different Models

To demonstrate that the current model combining multiple processing methods is optimal, this section conducts comparative experiments using the same sample data. Specifically, the models ICEEMEDAN-VMD-IIE-CNN, VMD-IIE-CNN-BiLSTM, and ICEEMDAN-IIE-CNN-BiLSTM are selected. The algorithm parameters and computational conditions are consistent with those described in Section 5.1.

Then, Figure 11 presents the confusion matrices for the three combined models mentioned above. Table 4 shows the fault state assessment accuracy for each model. The differences in classifiers have a minimal impact on overall accuracy, while the reconstruction (decomposition) methods for signals are crucial for the assessment. In summary, it is evident that the method reported in this paper effectively extracts fault features from samples, ensuring high accuracy in state evaluation.

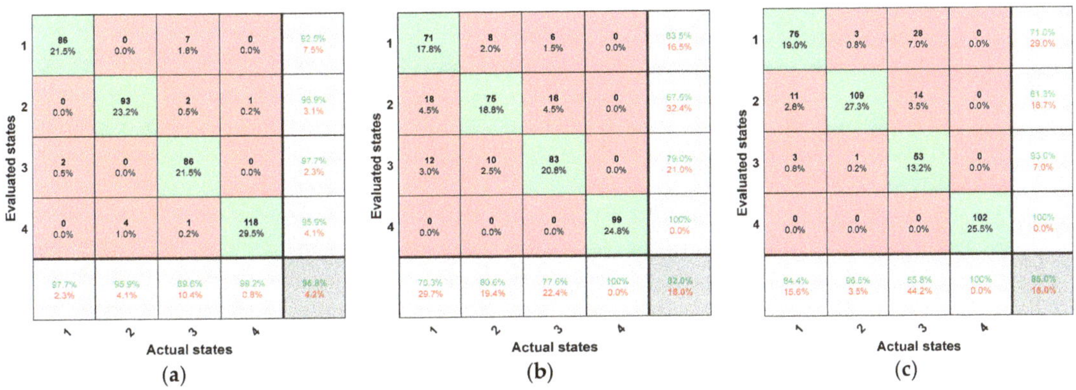

Figure 11. Accuracy of the evaluation results. (**a**) ICEEMEDAN-VMD-IIE-CNN, (**b**) VMD-IIE-CNN-BiLSTM, (**c**) ICEEMDAN-IIE-CNN-BiLSTM.

Table 4. Accuracy comparison of the evaluation results of various combinations.

	Reported Model	ICEEMEDAN-VMD-IIE-CNN	VMD-IIE-CNN-BiLSTM	ICEEMDAN-IIE-CNN-BiLSTM
Accuracy	99.25%	95.75%	82%	85%

Table 4 compares the proposed ICEEMEDAN-VMD-IIE-CNN with VMD-IIE-CNN-BiLSTM and ICEEMDAN-IIE-CNN-BiLSTM to highlight the contribution of ICEEMEDAN-VMD. According to this comment, we added the additional methods (SVM, GRNN, and DPRNN), and a detailed table of comparison with the state-of-the-art methods is tabulated in Table 5.

In the experiments, "×" indicates non-applicability, with decomposed signals used as feature inputs. The SVM parameters are set with a penalty factor of 0.1 and a radial basis function kernel; the GRNN parameters include a spread parameter of 0.5. The DPRNN parameters are configured as follows: seven layers, a stride of 1, ReLU activation function, a dropout rate of 0.3, a learning rate of 0.001, a batch size of 64, and the Adam optimizer. All other computational and sample conditions remain consistent.

According to the results, the impact of different classifiers on the evaluation outcomes is not significant when the signal reconstruction and feature extraction methods are the same. This is because the input features already exhibit good separability, as demonstrated by the TSNE results in Figure 5 in the revised manuscript.

Table 5. Accuracy comparison of the evaluation results corresponding to various models.

No.	Signal Reconstruction	Feature Selection	Classifier	Accuracy
1	ICEEMEDAN-VMD	IIE	MSCNN-BiLSTM	99.25%
2	ICEEMEDAN-VMD	IIE	CNN	82%
3	ICEEMDAN	IIE	CNN-BiLSTM	85%
4	ICEEMEDAN-VMD	×	MSCNN-BiLSTM	86.5%
5	ICEEMEDAN-VMD	IIE	SVM	93.75%
6	ICEEMEDAN-VMD	IIE	GRNN	95.25%
7	ICEEMEDAN-VMD	IIE	DPRNN	96.5%
8	ICEEMEDAN	IIE	DPRNN	86.25%
9	VMD	IIE	DPRNN	84.75%
10	ICEEMEDAN-VMD	×	DPRNN	88.5%

6. Conclusions

Given the faced shortcomings, this work proposed an intelligent strategy for fault diagnosis via the combination of signal reconstruction and deep learning, and the current findings lead to the following conclusions.

The strategy of combining ICEEMDAN and VMD enables adaptive signal decomposition, streamlining fault feature extraction through improved information entropy and avoiding the complexities of manual feature extraction while significantly enhancing computational efficiency. The developed fault classifier not only captures multi-scale spatial and temporal features but also integrates a channel attention mechanism, which improves the model's ability to select and extract fault features.

The comparative experimental results demonstrate that the proposed model exhibits strong generalization and versatility, maintaining high fault diagnosis accuracy (99.25%) under varying speeds and noisy conditions, outperforming comparison models. However, due to the low number of training and test samples, as well as the optimization of hyperparameters and algorithm architecture, further testing and research are needed to assess its applicability to other bearing faults and mixed fault scenarios.

Author Contributions: Conceptualization, W.L. and N.F.; methodology, C.Z.; software, C.Z. and M.L.; validation, X.Y.; formal analysis, L.M.; investigation, W.L. and X.P.; resources, N.F.; data curation, W.L. and N.F.; writing—original draft preparation, L.M.; writing—review and editing, M.L. and N.F.; project administration, W.L. All authors have read and agreed to the published version of the manuscript.

Funding: This research received no external funding.

Data Availability Statement: The data supporting the findings of this study are available from the corresponding author upon reasonable request due to the difficulty in obtaining experimental data.

Conflicts of Interest: The authors declare no conflicts of interest.

References

1. Hu, Z.X.; Wang, Y.; Ge, M.F.; Liu, J. Data-Driven Fault Diagnosis Method Based on Compressed Sensing and Improved Multi-Scale Network. *IEEE Trans. Ind. Electron.* **2020**, *67*, 3216–3225. [CrossRef]
2. Shao, H.D.; Xia, M.; Han, G.J.; Zhang, Y.; Wan, J. Intelligent Fault Diagnosis of Rotor-Bearing System under Varying Working Conditions with Modified Transfer CNN and Thermal Images. *IEEE Trans. Ind. Inform.* **2021**, *17*, 3488–3496. [CrossRef]
3. Knap, P.; Lalik, K.; Balazy, P. Boosted Convolutional Neural Network Algorithm for the Classification of the Bearing Fault from 1-D Raw Sensor Data. *Sensors* **2023**, *23*, 4295. [CrossRef]
4. Gao, L.; Li, X.K.; Yao, Y.C.; Wang, Y.; Yang, X.; Zhao, X.; Geng, D.; Li, Y.; Liu, L. A Modal Frequency Estimation Method of Non-Stationary Signal under Mass Time-Varying Condition Based on EMD Algorithm. *Appl. Sci.* **2022**, *12*, 8187. [CrossRef]
5. Guo, M.F.; Zeng, X.D.; Chen, D.Y.; Yang, N.C. Deep-Learning-Based Earth Fault Detection Using Continuous Wavelet Transform and Convolutional Neural Network in Resonant Grounding Distribution Systems. *IEEE Sens. J.* **2017**, *18*, 1291–1300. [CrossRef]
6. Zhang, Z.; Deng, A.; Wang, Z.; Li, J.; Zhao, H.; Yang, X. Wind Power Prediction Based on EMD-KPCA-BiLSTM-ATT Model. *Energies* **2024**, *17*, 2568. [CrossRef]
7. Wang, D.; Li, S.; Fu, X. Short-Term Power Load Forecasting Based on Secondary Cleaning and CNN-BILSTM-Attention. *Energies* **2024**, *17*, 4142. [CrossRef]

8. Zhang, Z.; Liu, X.; Zhang, X.; Yang, Z.; Yao, J. Carbon Price Forecasting Using Optimized Sliding Window Empirical Wavelet Transform and Gated Recurrent Unit Network to Mitigate Data Leakage. *Energies* **2024**, *17*, 4358. [CrossRef]
9. Huang, N.E.; Shen, Z.; Long, S.R.; Wu, M.C.; Shih, H.H.; Zheng, Q.; Yen, N.-C.; Tung, C.C.; Liu, H.H. The Empirical Mode Decomposition and the Hilbert Spectrum for Nonlinear and Non-Stationary Time Series Analysis. *Proc. Math. Phys. Eng. Sci.* **1998**, *903*, 903–995. [CrossRef]
10. Wu, Z.; Huang, N.E. Ensemble Empirical Mode Decomposition: A Noise Assisted Data Analysis Method. *Adv. Adapt. Data Anal.* **2009**, *1*, 1–41. [CrossRef]
11. Konstantin, D.; Dominique, Z. Variational Mode Decomposition. *IEEE Trans. Signal Process.* **2013**, *62*, 531–544.
12. Li, Z.; Chen, J.; Zi, Y.; Pan, J. Independence-Oriented VMD to Identify Fault Feature for Wheel Set Bearing Fault Diagnosis of High-Speed Locomotive. *Mech. Syst. Signal Process.* **2017**, *85*, 512–529. [CrossRef]
13. Ding, L.; Guo, H.; Bian, L. Convolutional Neural Networks Based on Resonance Demodulation of Vibration Signal for Rolling Bearing Fault Diagnosis in Permanent Magnet Synchronous Motors. *Energies* **2024**, *17*, 4334. [CrossRef]
14. Tama, B.A.; Vania, M.; Lee, S.; Lim, S. Recent advances in the application of deep learning for fault diagnosis of rotating machinery using vibration signals. *Artif. Intell. Rev.* **2023**, *56*, 4667–4709. [CrossRef]
15. Chen, X.H.; Yang, R.; Xue, Y.H.; Huang, M.; Ferrero, R.; Wang, Z. Deep Transfer Learning for Bearing Fault Diagnosis: A Systematic Review Since 2016. *IEEE Trans. Instrum. Meas.* **2023**, *72*, 3508221. [CrossRef]
16. Qiao, M.Y.; Yan, S.H.; Tang, X.X.; Xu, C. Deep Convolutional and LSTM Recurrent Neural Networks for Rolling Bearing Fault Diagnosis under Strong Noises and Variable Loads. *IEEE Access* **2020**, *8*, 66257–66269. [CrossRef]
17. Hao, S.J.; Ge, F.X.; Li, Y.M.; Jiang, J. Multisensor Bearing Fault Diagnosis Based on One-Dimensional Convolutional Long Short-Term Memory Networks. *Measurement* **2020**, *159*, 107802. [CrossRef]
18. Guo, Y.R.; Mao, J.; Zhao, M. Rolling Bearing Fault Diagnosis Method Based on Attention CNN and BiLSTM Network. *Neural Process. Lett.* **2023**, *55*, 3377–3410. [CrossRef]
19. Case Western Reserve University Bearing Data Center Website. 2013. Available online: https://engineering.case.edu/bearingdatacenter/ (accessed on 31 January 2023).
20. Li, Z.W.; Xu, H.Y.; Jiang, B.B.; Han, F. Wavelet Threshold Ultrasound Echo Signal Denoising Algorithm Based on CEEMDAN. *Electronics* **2023**, *12*, 3026. [CrossRef]
21. Pan, H.; Badawi, D.; Bassi, I.; Ozev, S.; Cetin, A.E. Detecting Anomaly in Chemical Sensors via L1-Kernel-Based Principal Component Analysis. *IEEE Sens. Lett.* **2022**, *6*, 7004304. [CrossRef]
22. Zivieri, R. Magnetic Skyrmions as Information Entropy Carriers. *IEEE Trans. Magn.* **2022**, *58*, 1500105. [CrossRef]
23. Fu, Y.; Zhou, K.; Zhu, G.; Li, Z.; Li, Y.; Meng, P.; Xu, Y.; Lu, L. A Partial Discharge Signal Separation Method Applicable for Various Sensors Based on Time–Frequency Feature Extraction of t-SNE. *IEEE Trans. Instrum. Meas.* **2024**, *73*, 3505609. [CrossRef]

energies

MDPI

Article

Optimization of Fault Current Limiter Reactance Based on Joint Simulation and Penalty Function-Constrained Algorithm

Jun Zhao [1,*], Chao Xing [1], Zhigang Zhang [1], Boyuan Liang [1], Lu Sun [1], Bin Wei [2], Weiqi Qin [2] and Shuguo Gao [1]

[1] Electric Power Research Institute, State Grid Hebei Electric Power Company, Shijiazhuang 050000, China; dyy_xingc@he.sgcc.com.cn (C.X.); zhangzg@he.sgcc.com.cn (Z.Z.); dyy_liangby@he.sgcc.com.cn (B.L.); dyy_sunl@he.sgcc.com.cn (L.S.); dyy_gaosg@he.sgcc.com.cn (S.G.)

[2] State Key Laboratory of Alternate Electrical Power System with Renewable Energy Sources, North China Electric Power University, Beijing 102206, China; 120202101040@ncepu.edu.cn (B.W.); 120222201389@ncepu.edu.cn (W.Q.)

* Correspondence: dyy_zhaoj@he.sgcc.com.cn

Abstract: This paper proposes a novel optimization method for fault current limiter (FCL) reactance configuration based on joint simulation and penalty function constraint optimization. By integrating MATLAB and ATP for joint simulation, the method accurately derives the constraint conditions of the objective optimization function, providing critical data support for the optimization process. To address the challenges of high computational complexity and solution difficulties in constrained optimization, the Penalty Function Method (PFM) is employed to transform the original constrained optimization problem into a standard unconstrained optimization problem, significantly reducing computational complexity and ensuring the feasibility of the solution. On this basis, the Gravitational Search Algorithm (GSA) is applied to compute the optimal reactance value. Through comparative analysis of engineering case studies, the superiority of the GSA over the Genetic Algorithm (GA) and Particle Swarm Optimization (PSO) in optimization performance is validated, further confirming the accuracy and efficiency of the proposed method. The results indicate that this method not only achieves precise calculation results but also significantly improves computational efficiency. Moreover, the integration of PFM and GSA demonstrates excellent robustness, providing reliable technical support for the optimized deployment of fast-switching fault current limiters in large-scale power grids.

Keywords: fault current limiter; optimal configuration of reactance value; joint simulation; penalty function method; gravitational search algorithm

1. Introduction

With the continuous expansion of China's power grid, the levels of short-circuit current across various voltage levels have been rising year by year, posing unprecedented challenges to the stability of the power system [1,2]. Taking the Hebei power grid as an example, calculations of the short-circuit withstand capacity of 220 kV transformers revealed that 73 transformers (accounting for 87%) had medium- and low-voltage side windings that did not meet the required standards. Among these, a significant number of transformers had major safety concerns on their medium-voltage side. Even more critically, in recent years, multiple incidents of medium-voltage side equipment damage due to short-circuit current impacts have occurred in the Hebei power grid, further highlighting the vulnerability of the system under short-circuit fault conditions.

Traditionally, solutions such as adjusting the grid structure or changing the operational mode are used to limit short-circuit currents [3]. However, these approaches require long construction periods, incur high costs, and may trigger a series of chain reactions, thereby reducing the overall stability and reliability of the power grid [4]. Additionally, installing fault current limiters (FCLs) on transmission lines can lead to increased reactive power losses during normal operation, affecting the power flow distribution [5]. Although installing FCLs at the transformer neutral points can limit short-circuit currents to some extent, unbalanced currents may still flow through the neutral point during operation, resulting in some power losses [6].

To address these challenges effectively, researchers have proposed using FCLs to limit short-circuit currents while avoiding energy losses from unbalanced currents during normal operation [7]. Recently, with advancements in fast-switching technology, fast-switching fault current limiters (FSFCLs) have gained widespread application. Compared to traditional FCLs, FSFCLs offer numerous advantages, including faster switching speeds, better economic performance, and higher reliability, making them an effective solution to the short-circuit current issue [8].

Existing studies have primarily focused on the FCL devices themselves and their applications on transmission lines [9,10]. For example, Gong Xianfu et al. [11] compared the effectiveness of installing short-circuit current limiting devices at transformer neutral points versus on transmission lines. Their results indicated that installing such devices near transformers not only reduces costs but also better suppresses short-circuit currents. Xia Shengguo et al. [12] conducted electromagnetic transient simulations to evaluate the effectiveness of FCLs with highly coupled split inductors on transmission lines. Professor Zhao Chengyong and colleagues [13] proposed a novel FCL topology for HVDC systems, which limits short-circuit current peaks by rapidly engaging energy-absorbing resistors and coordinating with DC circuit breakers. While these studies have focused on innovative devices and topologies, less attention has been given to selecting the optimal parameters and configuring the inductance.

In practical applications, traditional methods for calculating fault current limiter inductance typically involve setting an initial inductance value, followed by extensive short-circuit calculations for different installation positions, transformer operating modes, and fault types. These calculations provide short-circuit current values for various scenarios, ensuring that the currents do not exceed the permissible limits [14]. Additionally, the potential overvoltage across the neutral point insulation when installing a fast-switching fault current limiter must be calculated to ensure it remains below the transformer's insulation rating [15]. However, since inductance is inversely proportional to short-circuit current and directly proportional to overvoltage, this process often requires multiple trial calculations, making it time-consuming and inefficient. This inefficiency is especially problematic in large-scale grids, such as the Hebei power grid, where numerous variables and transformers need to be considered.

To enhance the efficiency of fault current limiting reactance configuration, this paper proposes an optimization method based on joint simulation and penalty function constraint optimization. The method sets the reactance value as the optimization objective and first utilizes joint simulation with MATLAB R2021b and ATP-EMTP to derive constraint conditions, providing precise data support for the optimization process. Subsequently, the PFM is introduced to transform the original constrained optimization problem into a standard unconstrained optimization problem, significantly reducing computational complexity and effectively addressing constraint issues. On this basis, the GSA is employed to compute the optimal reactance value. Through a comparative analysis of engineering cases, the superiority of the GSA over GA and PSO is demonstrated, further validating the accuracy and efficiency of the proposed method. The results indicate that the method not only delivers precise calculation results but also significantly improves computational efficiency. Moreover, the integration of PFM and GSA exhibits excellent robustness, making the method particularly suitable for the optimization and deployment of fast-switching fault current limiters in large-scale power grids. This optimization method effectively reduces computation time while providing reliable technical support for the widespread application of fast-switching fault current limiters in power systems.

2. Materials and Methods

2.1. Working Principle of FSFCL

The fast-switching fault current limiter consists of a current-limiting reactor, fast switch, overvoltage protection, controller, and current transformer (as shown in Figure 1) [8]. During normal operation, the fast switch remains closed, resulting in minimal device losses and no impact on the power grid. When the current transformer detects a fault current, the fast switch rapidly opens, and the current-limiting reactor is engaged to effectively control the fault current within the safe operating range of the transformer. After the fault is cleared, the fast switch closes again, ensuring the stable operation of the power grid [16]. The physical model of FSFCL is shown in Figure 2.

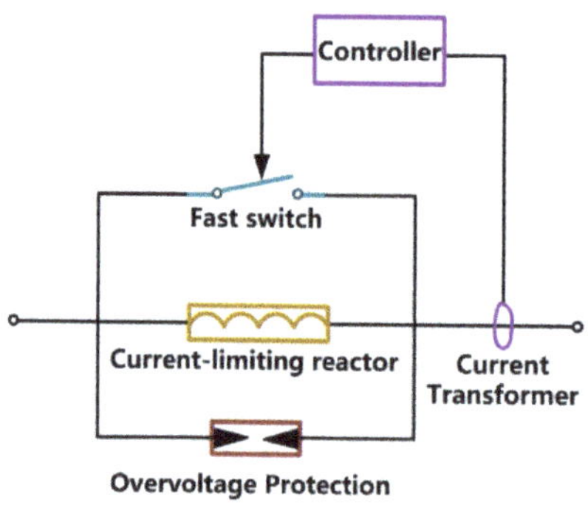

Figure 1. Schematic diagram of FSFCL structure.

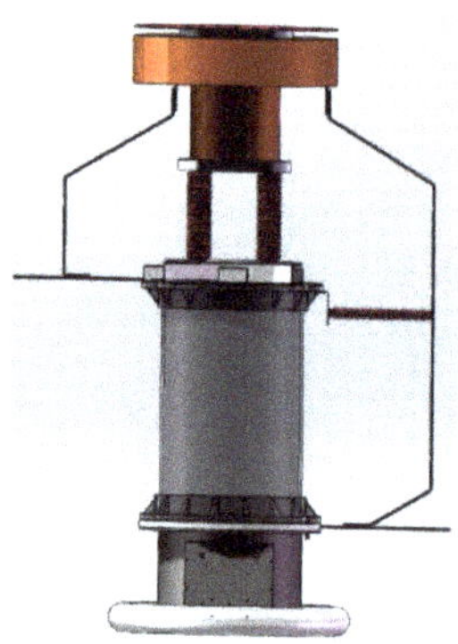

Figure 2. The physical model diagram of FSFCL.

2.2. Current-Limiting Reactor Inductance Optimization Model

2.2.1. Objective Solution Model for Current-Limiting Reactance Value

The cost of a fast-switching fault current limiter is primarily influenced by the equipment and installation costs. Since the fast-switching fault current limiter is mainly composed of a current-limiting reactor, fast switch, controller, and current transformer, most of the costs are difficult to reduce. The cost of the current-limiting reactor is positively correlated with the equivalent inductance of the reactor. Therefore, the objective function $\min f$ is set as the equivalent inductance value x of the current-limiting reactor in the fault current limiter.

$$\min f = x \tag{1}$$

The optimization of the fast-switching fault current limiter configuration must also satisfy the following constraint:

(1) Short-Circuit Current Constraint

To ensure the reliable operation of transformers, constraints should be imposed on short-circuit currents. According to the literature [17], when a short-circuit occurs in the nearby area, the short-circuit surge current through the windings increases significantly. If it exceeds 90% of the permissible short-circuit current, it may lead to the risk of transformer damage. Therefore, it is required that the maximum short-circuit current of each winding does not exceed 90% of the allowable short-circuit current. The short-circuit current constraint is thus set as follows:

$$\begin{cases} I_1(x) - \alpha I_B \leq 0 \\ I_2(x) - \alpha I_B \leq 0 \end{cases} \tag{2}$$

where $I_1(x)$ represents the maximum short-circuit current in each winding during a two-phase ground fault; $I_2(x)$ represents the maximum short-circuit current in each winding during a two-phase ground fault; α is a constant of 0.9, and I_B denotes the maximum allowable short-circuit current that can pass through each winding of the transformer.

(2) Overvoltage constraint

To prevent the overvoltage at the transformer neutral point from exceeding the insulation level and causing accidents after the current-limiting reactor is connected, constraints should be set for the neutral point overvoltage. According to the literature [18], the operating voltage of the equipment at the neutral point on the 110 kV side of the transformer is 72.5 kV. Its rated lightning impulse withstand voltage should be less than 325 kV, and the rated short-time power frequency withstand voltage should be less than 140 kV. Therefore, the following constraints are set for overvoltage:

$$\begin{cases} U_1(x) - v_1 < 0 \\ U_2(x) - v_2 < 0 \end{cases} \tag{3}$$

where $U_1(x)$ is the rated lightning impulse withstand voltage of the current-limiting reactor; $U_2(x)$ is the rated short-time power frequency withstand voltage of the current-limiting reactor; v_1 is the maximum lightning impulse withstand voltage, equal to 325 kV; v_2 is the maximum short-time power frequency withstand voltage, equal to 140 kV.

2.2.2. Objective Optimization Mode Based on the Penalty Function Method

The Penalty Function Method [19] adds a penalty term to the objective function with constraints, transforming a constrained optimization problem into an unconstrained one, thus avoiding the situation where no feasible solution can be found. In this study, the constraints, including short-circuit current and overvoltage, are converted into penalty function forms. These penalty terms are then incorporated into the objective function concerning the fault current limiter reactance value, resulting in a fitness function. This combined fitness function is used to efficiently solve the equivalent impedance value of the fault current limiter.

By introducing penalty functions k_i (i = 1, 2, 3, 4) and combining with Equations (1)–(3), the transformed conventional unconstrained optimization equation is as follows:

$$\min F(x,k_i) = f(x) + \sum_{i=1}^{4} k_i \cdot max(0, g_i(x)) \quad (4)$$

where $g_i(x)$ is derived from the constraint conditions as the penalty equation, as follows:

$$\begin{cases} g_1(x) = I_1(x) - \alpha I_B \\ g_2(x) = I_2(x) - \alpha I_B \\ g_3(x) = U_1(x) - v_1 \\ g_4(x) = U_2(x) - v_2 \end{cases} \quad (5)$$

3. Results and Discussion

3.1. *Joint Simulation Method Based on ATP and SIMULINK*

In this study, both Simulink and ATP-EMTP simulation tools were utilized, as each offers distinct advantages in specific application scenarios. Simulink excels in modeling low-frequency dynamics and control systems, making it particularly suitable for analyzing power frequency components in single-phase and two-phase ground fault currents (I_{k1}/I_{k2}), where steady-state or quasi-steady-state conditions dominate. Its electrical network models enable rapid solutions and facilitate the integration of control strategies, such as the timing and logic of fast-switch actions. However, for high-frequency electromagnetic transient analysis, Simulink may introduce larger errors or require more complex models.

In contrast, ATP-EMTP is highly precise for high-frequency phenomena, such as lightning overvoltage, due to its support for smaller simulation time steps and the use of distributed parameter models for transmission lines and BCT models for transformers. This allows ATP-EMTP to better capture high-frequency transient behaviors, including neutral point overvoltage (U_{TV}/U_{AC}). Although ATP-EMTP can also analyze short-circuit currents, it is less convenient for co-simulation with control systems and other tools like MATLAB, and it demands more computational resources for high-frequency transient handling.

By combining both tools, this study achieved a comprehensive and accurate approach to address various analysis tasks. This dual-tool strategy leverages the strengths of each—ensuring high-precision electromagnetic transient analysis with ATP-EMTP and flexible control system integration with Simulink—while mitigating the limitations associated with using a single tool.

3.1.1. Simulink Short-Circuit Current Simulation Calculation

To accurately assess the severe, short-circuit faults that may occur on the medium-voltage side under the most adverse conditions, short-circuit current simulation calculations need to be carried out. The simplified system topology is shown in Figure 3. In this paper, it is assumed that the transformer and its respective system sides all operate at maximum load, and the fault location is set at the near end of the medium-voltage side of the transformer [20].

Upon analysis, it was found that installing a current-limiting reactor at the high-voltage side neutral point does not significantly increase the zero-sequence impedance at the fault point, resulting in unsatisfactory fault current limitation. Additionally, the increased impedance on the high-voltage side can cause an increase in the zero-sequence fault current flowing into the low-voltage side windings during an unbalanced ground fault, potentially creating new safety hazards. Therefore, it is advisable to install the current-limiting reactor at the medium-voltage side's neutral point.

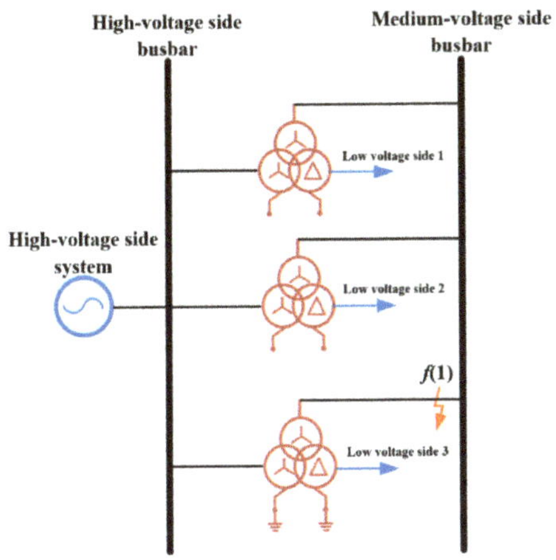

Figure 3. Power system topology diagram of a 220 kV substation.

The simulation model built in Simulink is shown in Figure 4, where the fault type can be set at the lightning symbol position in the figure. In this model, the input is the equivalent inductance value of the fast-switching fault current limiter. When the fault type is a single-phase ground fault, the short-circuit current value I_{k1} is calculated. When the fault type is a two-phase ground fault, the short-circuit current value I_{k2} is calculated.

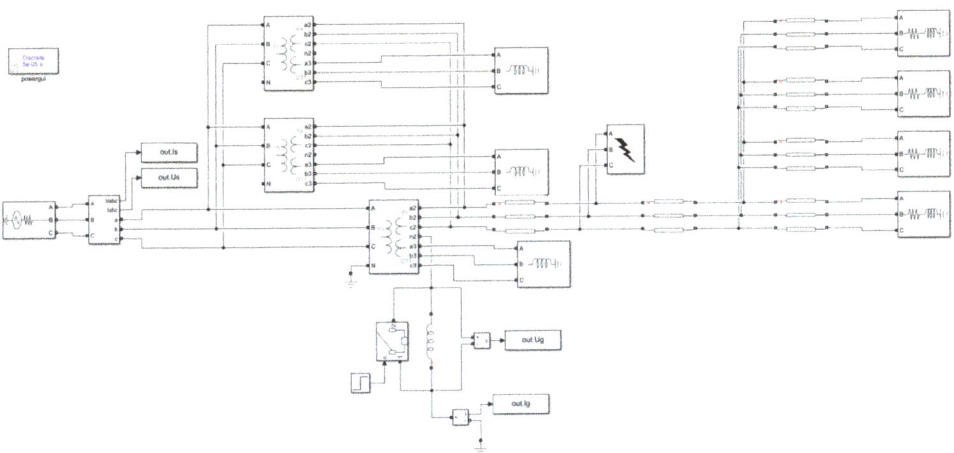

Figure 4. Simulink asymmetric short-circuit simulation model.

3.1.2. ATP-EMTP Overvoltage Simulation Calculation

During normal system operation, the neutral-point reactance only affects the zero-sequence loop and does not influence voltage or power flow. However, when an unbalanced short-circuit fault occurs on the line, the zero-sequence current flows through the neutral-point current-limiting reactor to the ground, which will cause the neutral-point voltage to rise [21]. If this voltage exceeds a certain threshold, it may lead to overvoltage at the transformer neutral point, surpassing the insulation withstand level, and potentially resulting in serious insulation failures. Therefore, before installing the fast-switching fault current limiter, overvoltage calculations must be performed considering the system wiring and transformer parameters to ensure the voltage remains within the insulation's withstand range.

In this paper, ATP-EMTP simulation software (The version number of the simulation-software used is GNU-Mingw32 ATP) is used to model the power frequency and transient overvoltage conditions after a current-limiting reactor is installed at the neutral point of a 220 kV substation's main transformer [22]. When building the simulation model, the surrounding power grid was simplified, and equivalent processing was applied, following the principle of maintaining power flow and node voltage unchanged to ensure the accuracy of the simulation data [23].

The simulation model is shown in Figure 5 and primarily consists of a transformer and the power grids on each side. The transformer model employs the SAT-saturated transformer model with a capacity of 120 MVA and a rated voltage of 220/121/11 kV. Both the excitation losses and short-circuit losses are set according to actual values. Three transformers operate in parallel on the high and medium-voltage sides, while the low-voltage side operates independently, with only one transformer grounded.

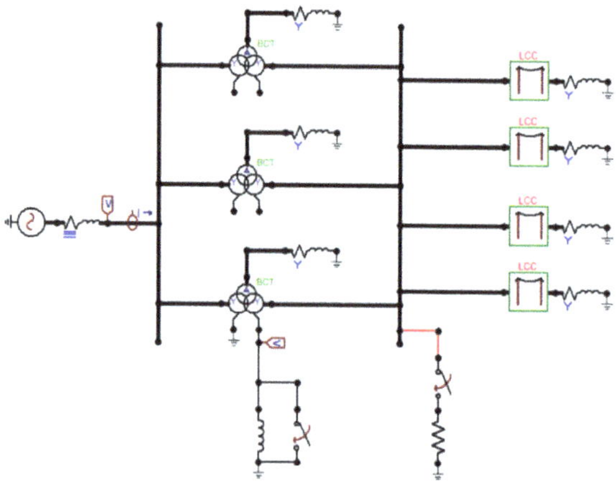

Figure 5. Simulation Circuit of a 220 kV Substation Power System.

For the power grids on each side:

1. High-Voltage Side: The 220 kV grid is modeled using a three-phase ideal voltage source and three-phase symmetrical parallel resistors for equivalence, with a peak phase voltage of 179.6 kV.

2. Medium-Voltage Side: The 10 kV grid consists of four transmission lines with lengths of 15.93 km, 3.35 km, 8.51 km, and 31.49 km, respectively. The ends of each transmission line are modeled using three-phase symmetrical Y-connected grounding resistors as equivalent loads.

3. Low-Voltage Side: The 11 kV grid is modeled using three-phase symmetrical Y-connected grounding resistors for equivalence.

In the simulation, the fault current limiter is simplified to a current-limiting reactor. Two types of asymmetrical fault conditions are set near the medium-voltage side outlet of the transformer: one is phase A grounded through a 0.01 Ω resistor, and the other is phases B and C grounded through a 0.01 Ω resistor.

The input to this model is the equivalent reactance value of the fast-acting switch-based fault current limiter, and the outputs are the rated transient withstand voltage U_1 and the rated short-time power frequency withstand voltage U_2.

3.1.3. Joint Simulation Method of ATP-EMTP and MATLAB

To achieve joint simulation between ATP-EMTP and MATLAB, it is necessary to control the parameters in ATP-EMTP through MATLAB, modifying the model to meet the overvoltage calculation requirements under different operating conditions. Additionally, the waveform data output from ATP-EMTP calculations is saved in the ".pl4" file, which is an encrypted file, making data reading difficult.

To adjust model parameters, MATLAB's text reading and modification functions are used to read and modify the model parameters and component states in the ATP-EMTP model configuration file.

To read the output results from ATP-EMTP, the STARTUP toolbar in the ATP-EMTP Tools toolbar is opened. In the "FMTPL4" option on the seventh tab, enter "$6 \times 10^{13.6}$" to convert the encrypted ".pl4" file output by ATP-EMTP into readable decimal data.

After solving these two issues, a joint simulation between MATLAB and ATP-EMTP was achieved by calling the ".tpbig" program in MATLAB. The detailed joint simulation process is shown in Figure 6.

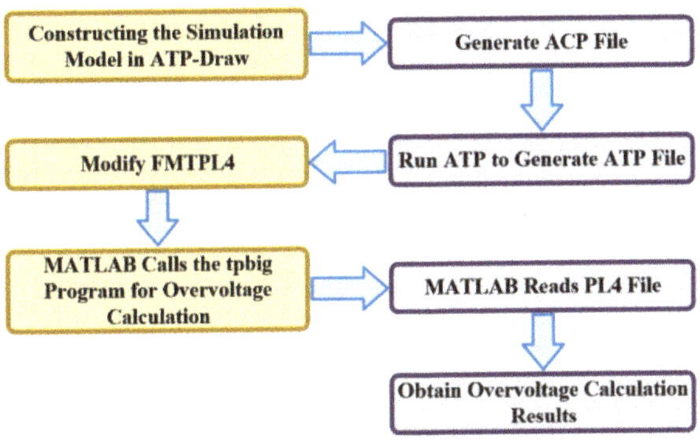

Figure 6. ATP-EMTP and MATLAB Joint Simulation Flowchart.

3.2. *Gravitational Search Algorithm-Based Optimization with Joint Simulation and Comparative Analysis*

3.2.1. Gravitational Search Algorithm-Based Optimization with Parameter Adaptation and Performance Comparison

However, the Penalty Function Method suffers from issues such as high computational cost and slow convergence speed [24]. Therefore, this study proposes using a GSA with adaptive dynamic adjustment of penalty parameters to avoid slow convergence caused by excessively large penalty parameters in the early stages, while ensuring effective penalization of constraint violations.

Gravitational Search Algorithm (GSA), a global optimization technique based on the law of gravity in nature, has been widely applied in various fields. As GSA has become a well-established method with numerous references available, this paper provides only a brief introduction to it [25–28].

The fundamental concept of GSA is to simulate the interaction between objects in a gravitational field to search for the optimal solution to a problem. In this algorithm, each solution is treated as an object, and the gravitational force between objects is determined by their mass and the distance between them. Larger mass objects exert a stronger gravitational pull on smaller ones, making their movement more pronounced. This allows larger objects to effectively guide smaller ones toward the global optimum.

During initialization, a set of solutions is randomly generated, with each solution's mass proportional to its fitness. For each pair of objects i and j, the gravitational force $F_{ij}(t)$ at the t-th iteration is calculated using the gravitational formula:

$$F_{ij}(t) = G\frac{M_i(t) \times M_j(t)}{{R_{ij}}^2(t)} \tag{6}$$

where G is the gravitational constant, $M_i(t)$ and $M_j(t)$ are the masses of objects i and j at the t-th iteration, and $R_{ij}(t)$ is the distance between them at the same iteration. The size of the gravitational force determines the degree of attraction between the objects, thereby influencing their movement direction and speed.

During each iteration, the positions of the objects are updated according to the gravitational forces, following the update formula:

$$x_i(t+1) = x_i(t) + \sum_{j=1}^{N} F_{ij}\cdot(x_j(t) - x_i(t)) \tag{7}$$

where $x_i(t)$ is the position of object i at the t-th iteration, and $x_j(t)$ is the position of object j. The gravitational force F_{ij} between objects i and j affects their movement toward each other. The parameter N denotes the number of particles in the population. With each iteration, the objects gradually move toward the optimal solution. As the iteration count increases, the gravitational force decreases, causing the objects' velocities to reduce until they converge to the optimal solution or meet the stopping criterion.

To validate the superiority of GSA over conventional optimization methods in FCL reactance configuration, we conducted comparative analyses with Genetic Algorithm (GA) and Particle Swarm Optimization (PSO) under two scenarios.

1. Single Reactance Optimization

Based on the practical validation provided in the case study in Section 3.3, which focuses on the Hebei power grid, a comparative analysis of the GSA, GA, and PSO was conducted. Using the example of a FCL installed at the neutral point of a 220 kV transformer, the following parameter settings were applied: for the PSO algorithm, the inertia weight is set to 0.6, and the learning factors are set as c1 = 1.7 and c2 = 1.5 [29]; for the GA, the crossover probability is set to 0.8, and the mutation probability is set to 0.1 [30]. The comparison results are summarized in the Table 1.

Table 1. Comparison of the Convergence Performance of Different Optimization Methods.

Algorithm	Average Convergence Iterations	Optimal Solution x (mH)	Solution Std. Dev σ (mH)	Computation Time (s)
GSA	83	15.9	<0.01	948.1
GA	33	15.9	<0.01	363.7
PSO	134	15.9	<0.01	1436.8

The results show that in the optimization of a single current-limiting reactance value, the convergence rate follows GA > GSA > PSO, with all three algorithms stably converging to the same optimal solution (15.9 mH) and a standard deviation below 0.01 mH. Additionally, the results indicate that computation time is primarily influenced by the number of iterations, as each algorithm iteration takes less than 0.1 s, while the simulation step for each iteration exceeds 10 s. Although GA exhibits the fastest convergence, practical applications prioritize robust global search capabilities to ensure stable identification of the global optimal solution. In real-world engineering scenarios, where multiple FSFCLs are installed at various points, the optimization of reactance values evolves into a multimodal problem with numerous local optima, making GSA's global search advantage critically important.

2. Multi-Reactance Optimization

Consider the following scenario: three FSFCLs are installed at the neutral points of a 220 kV substation, and one FSFCL is installed on each of five feeder lines. The objective function in this case can be simply set as:

$$\min f = \sum_{j=1}^{8} \lambda_j x_j \tag{8}$$

where x_j represents the current-limiting reactance value of each FSFCL, and λ_j represents the influence weight of each reactance value. In this scenario, the weights λ_j are assumed to be set as 1, 0.5, 2, 1, 1.5, 2, 1, and 0.5, respectively.

The constraints are assumed to remain unchanged. At this point, different combinations of reactance values may lead to similar objective function values, meaning the objective function has multiple local optima. After running the program for 100 iterations, the results are summarized in Table 2.

Table 2. The Global Optimal Solution Hit Rate of Different Optimization Methods.

Algorithm	Global Optimal Solution Hit Rate
GSA	92%
GA	78%
PSO	67%

Through the above simple case, it can be observed that the GSA demonstrates superior global search capabilities compared to the other two methods, making it more suitable for practical engineering applications in the optimization and selection of current-limiting reactance values.

3.2.2. Gravitational Search Algorithm Key Parameter Settings

The main advantage of GSA lies in its strong global search capability, which helps in preventing entrapment in local optima. This feature makes it especially suitable for high-dimensional, complex optimization problems. By utilizing gravitational interactions between objects, GSA promotes comprehensive exploration of the solution space and demonstrates excellent convergence and optimization performance.

To further enhance the application of GSA in optimizing FCL reactance configurations, several key parameters are carefully designed as follows:

1. GSA Parameter Design

(1) Gravitational Constant Evolution

The gravitational constant G is dynamically adjusted using an exponential decay function to balance exploration and exploitation:

$$G(t) = G_0 \cdot e^{-\alpha (t/T)^{\beta}} \tag{9}$$

where β = 0.7; G_0 = 100, and α = 20, are empirical coefficients derived from power grid optimization studies [31].

(2) Initialization Strategy

1) Position Initialization

Adopting the asymmetric sampling method from [32]:

$$x_i(0) = x_{\min} + \left[(rand + 0.1i)^{1.5}\right] \cdot (x_{\max} - x_{\min}) \tag{10}$$

2) Mass Initialization

Fuzzy-transformed fitness values characterize object masses:

$$M_i(0) = \frac{\exp(-F_i/\sigma)}{\sum_{j=1}^{N} \exp(-F_j/\sigma)} \tag{11}$$

where F_i represents the fitness value of the i-th object, σ is the standard deviation of the fitness values (used to normalize the distribution), and N is the population size.

3) Velocity Initialization:

Initial velocities $v_i(0) = 0$ minimize early-stage divergence risks.

(3) Population Parameters

1) Population Size

The population size N is as follows [33]:

$$N \geq 10 \cdot n \tag{12}$$

where n is the problem dimension ($n = 1$ in this study).

However, considering practical application scenarios where the selection of current-limiting reactance values for multiple FSFCLs may be required, the number of devices typically does not exceed ten. Taking into account convergence efficiency, solution accuracy, and computational cost, a population size of $N = 100$ was ultimately chosen.

2) Maximum Iterations

Based on the convergence analysis in [34], the maximum number of iterations, D, should meet the following condition:

$$P(D) \geq 99.8\% \tag{13}$$

where $P(D)$ denotes the probability of convergence within D iterations.

During the actual iterative computation process, the iteration count required for solution convergence occasionally exceeds 100. Therefore, the maximum number of iterations is set to 200 in this study to ensure a sufficient convergence probability while maintaining a certain margin.

During the iterations, the penalty factor is dynamically adjusted based on the current constraint violations. The penalty factor in this study is initially set to 1 and increases proportionally with each iteration, intensifying the penalty for constraint violations. This dynamic adjustment mechanism ensures that the algorithm prioritizes solutions that satisfy the constraints throughout the search process, guiding the algorithm toward feasible solutions.

When the objective function value changes very little over several iterations and satisfies the convergence criteria, it indicates that the algorithm has found a solution close to the optimal one. At this point, the algorithm can be terminated. Through this iterative process, the optimal reactance value for the fault current limiter (FCL) is determined.

The selection of $\Delta k = 1.1$ in this study is based on the dynamic adjustment theory [24], which employs a linearly increasing penalty coefficient to prevent insufficient constraint weighting during the early iterations. This incremental strategy has been validated as effective in the IEEE CEC 2017 benchmark tests.

In addition to the penalty factor's incremental adjustment, the selection of the initial penalty parameter k_{n0} requires careful consideration. A balance must be achieved between its impact on constraint satisfaction and convergence performance. If k_{n0} is set too small, the algorithm may fail to adequately penalize constraint violations, leading to unacceptable errors. Conversely, if k_{n0} is set too large, it may increase computational complexity and reduce algorithm efficiency. Therefore, the selection of k_{n0} must be approached with caution in practical applications.

For the computational scenario in this study, both current and voltage constraints are critical. Violating these constraints would lead to unacceptable outcomes, making it essential to set k_{n0} sufficiently large to strictly enforce the constraints while avoiding overly small values. To evaluate the impact of different initial penalty parameter values, in the context of the case study in Section 3.3 of this paper, an analysis was conducted using k_{n0} values of 0.001, 0.01, 0.1, 1, 10, and 100. The results of this analysis are shown in Table 3.

Table 3. Comparison of convergence performance under different k_{n0} values.

k_{n0}	Average Convergence Iterations	Optimal Solution x (mH)
0.001	82	8.9
0.01	83	14.1
0.1	83	15.7
1	83	15.9
10	82	15.9
100	83	15.9

As shown in Table 3, all parameter settings achieve convergence with nearly identical convergence iterations. However, when k_{n0} is relatively small (e.g., 0.001 or 0.01), the resulting optimal reactance value fails to meet the constraint conditions and shows significant deviation from the expected value. As k_{n0} increases, the optimization results gradually stabilize, and the constraints are strictly satisfied.

Considering both convergence accuracy and computational efficiency, this study selects $k_{n0} = 1$ as the initial penalty parameter. This value ensures that the constraints are strictly enforced while maintaining computational efficiency and solution accuracy.

The overall calculation process proposed in this paper is shown in Figure 7.

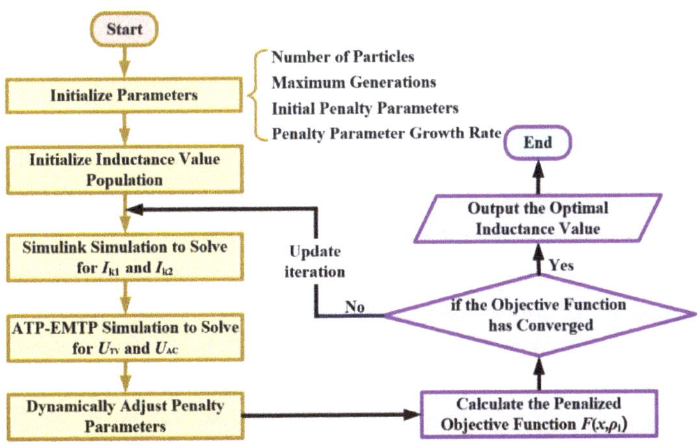

Figure 7. Flowchart for Adaptive Dynamic Adjustment of Penalty Parameters and Solution of Objective Function.

3.3. Case Study Analysis

This paper takes the No. 1 main transformer of a 220 kV substation in the Hebei power grid as an example to analyze the situation of asymmetric short circuits occurring on the medium-voltage side of the transformer. The high-voltage side of the transformer is independently powered, and the high- and medium-voltage sides operate in parallel with two other transformers that have nearly identical parameters, while the low-voltage side operates independently.

The basic parameters of the transformer are as follows: rated voltage of 220/121/11 kV, rated capacity of 120/120/60 MVA, short-circuit impedance of 14.1%, 24.0%, and 7.32%, connection group YNyn0D11, and the transformer core is a three-phase five-column type.

The current-limiting target for the windings on each side of the transformer is set to 1.94/2.85/17.90 kA.

Using the optimization configuration method proposed in this paper, the optimal inductance value calculated is 15.9 mH. The variation and convergence process of the current-limiting reactance values obtained through iterations is shown in Figure 8.

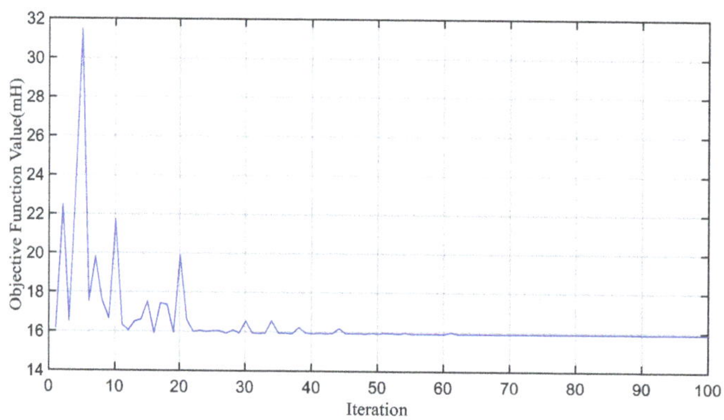

Figure 8. Convergence Curve of the Objective Function Using the GSA.

Based on the convergence curve, the current-limiting reactor value shows significant oscillations during the first 20 iterations, ranging from approximately 16 mH to 30 mH, indicating that the GSA is still in the global search phase and has not yet converged. Between the 20th and 45th iterations, the oscillations decrease, and the reactor value gradually stabilizes around 16mH, reflecting partial convergence. From the 50th iteration onward, the reactor value becomes increasingly stable, achieving full convergence near the 80th iteration and locating a region close to the optimal solution.

At this reactance value, the short-circuit current values for each winding under different fault conditions, as well as the maximum transient overvoltage and maximum power frequency steady-state overvoltage at the neutral point of the medium-voltage side, are presented in Table 4.

Table 4. Calculation Results under Various Fault Conditions.

Fault Conditions	High-Voltage Side Winding Current (kA)	Medium-Voltage Side Winding Current (kA)	Low-Voltage Side Winding Current (kA)
Single-phase ground fault	0.92	2.78	7.05
Two-phase ground fault	1.80	2.85	4.54

In both fault scenarios, the current-limiting targets are achieved, and the short-circuit current constraints are met. Additionally, the maximum transient overvoltage on the medium-voltage side neutral point is 74.6 kV, and the maximum power–frequency steady-state overvoltage is 49.8 kV. Both values are lower than the insulation level of the 110 kV side neutral point equipment, thus satisfying the overvoltage constraint requirements.

Based on the comprehensive analysis presented above, this paper ultimately ascertains the optimal limiting reactance value for the equipment to be 15.9 mH. A corresponding fault current limiter has been fabricated accordingly. The subsequent step entails preparing for commissioning and field application. The specific physical prototype is depicted in the Figures 9 and 10.

Figure 9. Photograph of the current-limiting reactor.

Figure 10. Overall physical image of the fault current limiter.

4. Conclusions

This paper proposes a novel optimization method for FSFCL reactance configuration based on joint simulation and penalty function constraint optimization. By integrating MATLAB and ATP-EMTP for joint simulation, the method effectively derives the constraint conditions of the objective optimization function, providing critical data support for the optimization process. To address the challenges of high computational complexity and solution difficulties in constrained optimization problems, the PFM is employed to transform the constrained optimization problem into a standard unconstrained optimization problem, significantly reducing computational complexity and ensuring solution feasibility. Subsequently, the GSA is applied to compute the optimal reactance value. Through a comparative analysis based on engineering case studies, the superiority of the GSA over GA and PSO is validated, further confirming the accuracy and efficiency of the proposed method. The results demonstrate that the method not only achieves precise parameter calculations but also significantly enhances computational efficiency. By combining PFM and GSA, the proposed approach exhibits excellent robustness, providing reliable technical support for the application and deployment of fast-switching fault current limiters in large-scale power grids.

Author Contributions: Supervision, J.Z. and S.G.; Investigation, Z.Z.; Writing—Original Draft Preparation, B.W.; Writing—Review and Editing, C.X. and L.S.; Project Administration, B.L. and W.Q. All authors have read and agreed to the published version of the manuscript.

Funding: This research was funded by the science and technology project of Hebei Electric Power Company of state grid (kj2023-013, research on short-circuit damage prevention technology of medium-voltage side winding of transformers based on large capacity fast switch).

Data Availability Statement: Data are contained within the article.

Conflicts of Interest: Authors Jun Zhao, Chao Xing, Zhigang Zhang, Boyuan Liang, Lu Sun, and Shuguo Gao were employed by the Electric Power Research Institute of State Grid Hebei Electric Power Company. The remaining authors declare that the research was conducted in the absence of any commercial or financial relationships that could be construed as a potential conflict of interest.

References

1. Yu, Y.; Wei, M.; Lyu, P.; Yang, G.; Wang, C. A practical modeling method for new energy stations applicable to asymmetric faul short-circuit calculation. *Power Grid Technol.* **2024**, *48*, 4788–4795.
2. Liu, Y. Analysis and treatment of low-voltage short-circuit faults in 220 kV main transformers. *Hebei Electr. Power Technol.* **2017** *36*, 53–55.
3. Zheng, S.; Liu, H.; Zhu, S. Two-step optimization method for limiting measures of three-phase and single-phase short circuits *Electr. Meas. Instrum.* **2019**, *56*, 70–75+81.
4. Han, B.; Chen, W.; Chang, N.; Jin, Y.; Qie, X.; Han, Y.; Zhu, B.; Tan, S.; Yu, Y.; Zhang, J.; et al. Research on short-circuit curren limitation method based on system topology dynamic adjustment. *Grid Technol.* **2021**, *45*, 1158–1166.
5. Zhao, E.; Han, Y.; Liu, Y.; Zalhaf, A.S.; Wang, C.; Yang, P. Feasibility analysis of neutral grounding by small reactor of HVDC converter transformer. *Energy Rep.* **2022**, *8* (Suppl. S1), 392–399. [CrossRef]
6. Chen, T.; Li, B.; Han, Y.; Li, Q.; Wang, J.; Jin, J. Research on fault line selection of small current grounding system considering zerc sequence unbalanced current. *Electr. Porcelain Light. Arrester* **2022**, 124–130.
7. Liu, J.; Yuan, J.; Zhou, H.; Li, X.; Mo, Z. Parameter design of short-circuit fault limiter considering the transient component o short-circuit current in high-voltage AC systems. *High Volt. Technol.* **2024**, *50*, 3769–3785.
8. Wang, W.; Zhao, Y.; Xu, Y.; Lv, W.; Yang, B.; Fang, T. Canister-type rapid switch for 500 kV fast switching fault current limite *High Volt. Technol.* **2023**, *49*, 803–811. [CrossRef]
9. Im, I.G.; Choi, H.S.; Jung, B.I. Limiting characteristics of the superconducting fault current limiter applied to the neutral line o conventional transformer. *Phys. C Supercond.* **2013**, *494*, 339–343. [CrossRef]

10. Sahebi, A.; Samet, H. Discrimination between internal fault and magnetizing inrush currents of power transformers in the presence of a superconducting fault current limiter applied to the neutral point. *IET Sci. Meas. Technol.* **2016**, *10*, 537–544. [CrossRef]
11. Gong, X.; Gao, C.; Long, Z.; Lin, Y.; Xu, L. Principle of asymmetric short-circuit current limitation by adding a small reactance at the neutral point of 500kV auto-transformer. *Electr. Power Constr.* **2013**, *34*, 56–60.
12. Xu, M. Research on Modeling and Application of Complex High Coupling Split Reactor for Fault Current Limiter. Master's Thesis, Huazhong University of Science and Technology, Wuhan, China, 2020.
13. Han, N.; Jia, X.; Zhao, X.; Xu, J.; Zhao, C. A new topology of hybrid DC fault current limiter. *Proc. Chin. Soc. Electr. Eng.* **2019**, *39*, 1647–1658+1861.
14. Hong, J.; Guan, Y.; Xu, G. Selection of fault current limiter installation position and parameter calculation in large power grids. *High Volt. Appar.* **2010**, *46*, 62–66.
15. Hu, J.; Li, X.; Yang, Y.; Bao, X.; Li, J. Analysis of neutral grounding overvoltage of 220 kV transformer. *Electr. Porcelain Light. Arrester* **2019**, 152–158.
16. Xu, D. Research on Fault Current Limiter Based on Fast Switch. Master's Thesis, Dalian University of Technology, Dalian, China, 2021.
17. *Q/GDW169-2008*; Oil-immersed Transformers (Reactors) Condition Evaluation Guide. State Grid Corporation of China: Beijing, China, 2008.
18. *GB 311.1-2012*; Insulation Coordination Part 1: Definitions, Principles and Rules. Standardization Administration of China: Beijing, China, 2012.
19. Jia, L.; Li, W.; Zhao, C. Numerical optimization design of vehicle cooling fan based on penalty function method. *Mech. Des. Manuf.* **2020**, 201–205+210.
20. Zhang, B.; Zhang, W.; Guo, S.; Wang, X.; Li, T. Simulation calculation of transformer differential current containing non-periodic components in external short circuits. *Hebei Electr. Power Technol.* **2023**, *42*, 82–85+94.
21. Zhao, H.; Zhang, X.; Cao, W.; Cheng, Z.; Gao, S. Study on overvoltage protection of ultra-high voltage transformer neutral point grounded through a small reactance. *Insul. Surge Arresters* **2022**, 109–115.
22. Zhang, W.; Ding, Q.; Hu, G.; Qiao, Z.; Ma, X. Research on lightning overvoltage protection scheme of distribution transformer based on ATP-EMTP simulation analysis. *Electr. Porcelain Light. Arrester* **2020**, 86–92.
23. Kang, S.; Feng, W.; Qin, Z.; Jia, H. Overvoltage analysis of fast power supply restoration for transformers with ungrounded neutral points after faults. *Hebei Electr. Power Technol.* **2019**, *38*, 5–8.
24. Si, C.; Lan, T.; Hu, J.; Wang, L.; Wu, Q. Discussion on the penalty coefficient in the penalty function. *Control. Decis.* **2014**, *29*, 1707–1710.
25. Aditya, N.; Mahapatra, S.S. An adaptive gravitational search algorithm for optimizing mechanical engineering design and machining problems. *Eng. Appl. Artif. Intell.* **2024**, *138*, 109298. [CrossRef]
26. Roy, A.; Verma, V.; Gampa, S.R.; Bansal, R.C. Planning of distribution system considering residential rooftop photovoltaic systems, distributed generations and shunt capacitors using gravitational search algorithm. *Comput. Electr. Eng.* **2023**, *111*, 108960. [CrossRef]
27. Shukla, A.; Momoh, J.A. Pseudo inspired gravitational search algorithm for optimal sizing of grid with integrated renewable energy and energy storage. *J. Energy Storage* **2021**, *38*, 102565. [CrossRef]
28. Omkar, S.; Arabinda, G.; Kumar, A.R. Estimation of parameters of one-diode and two-diode photovoltaic models: A chaotic gravitational search algorithm-based approach. *Energy Sources Part A Recovery Util. Environ. Eff.* **2023**, *45*, 5938–5956.
29. Marini, F.; Walczak, B. Particle swarm optimization (PSO): A tutorial. *Chemom. Intell. Lab. Syst.* **2015**, *149*, 153–165. [CrossRef]
30. Coello, C.A. An updated survey of GA-based multiobjective optimization techniques. *ACM Comput. Surv. (CSUR)* **2000**, *32*, 109–143. [CrossRef]
31. Zhang, X.; Liu, B.; Wang, Y. Day-ahead reactive power optimization scheduling strategy for active distribution networks considering operating condition prediction errors. *J. North China Electr. Power Univ. (Nat. Sci. Ed.)* **2024**, *51*, 31–40.
32. Rashedi, E.; Nezamabadi-Pour, H.; Saryazdi, S. GSA: A gravitational search algorithm. *Inf. Sci.* **2009**, *179*, 2232–2248. [CrossRef]
33. Sareni, B.; Krahenbuhl, L. Fitness sharing and niching methods revisited. *IEEE Trans. Evol. Comput.* **1998**, *2*, 97–106. [CrossRef]
34. Deb, K.; Agrawal, R. Simulated binary crossover for continuous search space. *Complex Syst.* **1995**, *9*, 115–148.

energies

MDPI

Article

Optical Detection and Cluster Analysis of Metal-Particle-Triggered Alternating Current Optical Partial Discharge in SF_6

Hanhua Luo [1], Yan Liu [2], Chong Guo [2] and Zuodong Liang [1,*]

[1] School of Electrical Engineering, Tsinghua University, Beijing 100084, China; lhh22@mails.tsinghua.edu.cn
[2] China Electric Power Research Institute, Beijing 100192, China; liuyan1@epri.sgcc.com.cn (Y.L.); guochong@epri.sgcc.com.cn (C.G.)
* Correspondence: lzd21@mails.tsinghua.edu.cn

Abstract: Accurately detecting defect-induced photon emissions enables early defect detection and characterization. To address this, a defect evolution state recognition model based on phase-resolved photon counting and dimensionality reduction calculations is proposed under alternating current (AC) excitation. Initially, photon information from protruding metal defects simulated using needle–plane electrodes during partial discharge (PD) evolution is analyzed in SF_6. Subsequently, phase-resolved photon counting (PRPC) techniques and statistical analysis are employed to extract feature parameters for quantitative characterization of defect-induced photon responses. Finally, a t-distributed stochastic neighbor embedding (t-SNE) dimensionality reduction analysis is utilized to establish criteria for categorizing defect evolution states. The findings reveal that metal-particle-triggered optical PRPC maintains the obvious polarity effect, and the entire evolution of the discharge can be divided into three processes. These research findings are expected to advance the accurate assessment of operational risks in gas-insulated systems.

Keywords: electroluminescence; epoxy resin insulation; defect detection; photon counting; phase-resolved photon counting

1. Introduction

Gas-insulated equipment (GIE) is widely utilized in ultra-high-voltage transmission networks in China due to its excellent properties [1,2]. However, various defects (gas gap, scratch, metal particles, etc.) unavoidably exist inside gas-insulated equipment and evolve under prolonged exposure to strong electric fields, leading to discharge faults [3].

To characterize these defects, methods based on PD signals have been extensively reported [4,5]. Research indicates that defect types and sizes influence discharge characteristics, reflected in their phase-resolved partial discharge (PRPD) patterns [4,5]. Statistical analysis of PRPD patterns enables the quantitative analysis of defect states. Nevertheless, conventional PD detection methods are prone to false alarms and misses due to noise in field environments. In contrast, optical detection offers inherent advantages such as immunity to electromagnetic interference and equipment vibrations, enhanced by the fully enclosed structure of GIE [6]. Given this, the optical-based detection method is regarded as a potential tool for detecting defects in GIE.

Reviewing the existing research, optical-based studies are always performed under either direct current (DC) or AC conditions. A quite different measurement result can be observed when the applied voltage is in different forms [7]. It is revealed that the optical signal can be more readily collected under AC excitation compared with DC conditions. The reason for the above phenomena is attributed to the difference in the charge transportation mechanism [8]. In particular, two types of luminous phenomena can be observed in the entire AC phase, but the luminous phenomenon can be observed at a higher voltage that triggers excitation or ionization and even the discharge of the charges inside the insulation materials. Consequently, research under AC excitation deserves more attention. F. Baudoin et al. established a bipolar charge transport model using polyethylene's electroluminescence to explain the microscopic principles of charge transport and luminescence in dry nitrogen [9]. Bamji identified the electroluminescence onset voltage as a threshold for polymer degradation, marking the onset of insulation deterioration in a vacuum [10]. Additionally, B. Qiao highlighted that optical emission from PD is typically orders of magnitude higher than electroluminescence in liquid nitrogen [11]. Thus, electroluminescence (EL) marks the initial stage of PD development, offering high sensitivity for detecting micro-defects and early warnings based on accurate photon emission measurements [12]. Given that SF_6 is currently the most widely used insulating gas in GIE, research by Ren Ming et al. on optical measurements in SF_6 across different wavelengths and discharge types demonstrated significant spectral differences in the spectrum for various defects [13,14]. Moreover, reference [15,16], which studied photon radiation, quantitatively explored electroluminescence characteristics induced by defects, showing significantly lower onset voltages and intensity correlating with defect severity compared to PD. The findings also revealed that one promising optical approach is phase-resolved photon counting (PRPC), which enables the detailed characterization of defect-induced photon emissions.

However, existing studies primarily focus on photon responses under different defect conditions, with limited reporting on the evolution of photon features during defect evolution. Given this, to explore photon emission information during defect evolution for the quantitative characterization of defects, this paper explores PRPC techniques to establish a more accurate defect assessment framework. Initially, it investigates the PRPC features of protruding metal defects under increasing excitation voltages, proposing multi-physical features for a quantitative description of the PRPC pattern through statistical analysis. Furthermore, it employs t-SNE [17] dimensionality reduction to analyze the different stages of defect evolution. The findings provide insights into defect severity classification and operational risk assessment.

2. Experimental Platform and Measurement Results

2.1. Experimental Platform

An aluminum needle (length is 10 mm, and tip curvature diameter is 1 mm) was utilized to simulate a metal protrusion in GIE equipment. The distance between the needle tip and the ground electrode was fixed at 7 mm. The simulation electrode and measurement circuit are plotted in Figure 1, in which the applied voltage was a growing 50 Hz AC voltage, and the released photons were recorded by a photon counting sensor (H8259-1, Hamamatsu Photon, Hamamatsu, Japan) with a resolution of less than 35 ns, a dark count of less than 80 s^{-1}, and an available spectral region of 185 to 850 nm. The counting unit was used for transforming the output pulses into the counting results as well as providing the voltage. Then, a data recorder (Pico Scope 2000, Pico Technology, St. Neots, UK) was used to trigger the synchronizing measurements and for the visualization of the measurement results.

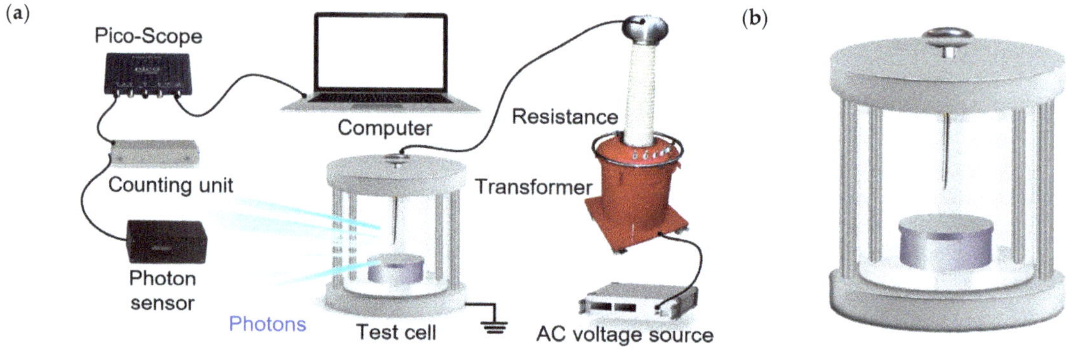

Figure 1. (**a**) Measurement platform and circuit for photon counting measurement, (**b**) electrode model for simulating a metal protrusion defect in GIE equipment.

Then, a growing AC voltage with a range from 4.5 to 11.5 kV was applied, since the inception of the photons occurred at around 4.0 kV. As per to [15], to observe more photons compared with a higher pressure, all the performed tests were carried out under 0.1 Mpa SF_6. The data record was divided into 15 steps, and each step lasted 20 min (the first 10 min was used to ensure stable luminescence) to collect the photon counting data with 200 AC cycles at room temperature. The scheme for applying the voltage is shown in Figure 2.

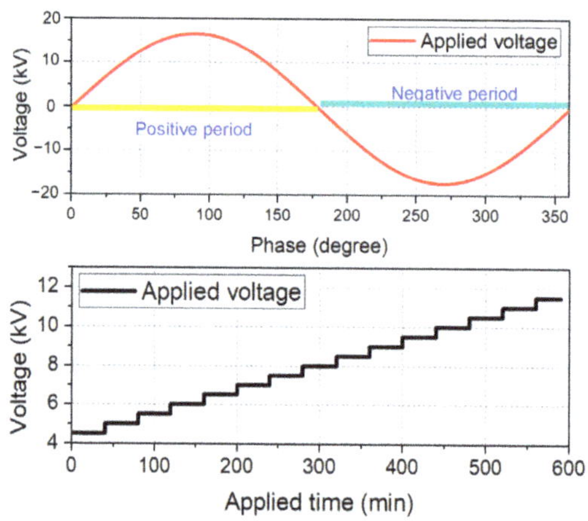

Figure 2. The phase of the applied voltage and the scheme for increasing the voltage.

2.2. Measurement Results

The metal protrusion shown in Figure 1b was tested under 15 voltages, and a total of 9 voltages were selected and used for plotting the PRPC pattern shown in Figure 3. Specifically, the PRPC pattern was generated by correlating the time-domain photon counting pulses with the phase angle of the applied AC voltage. High-frequency sensors captured the pulses, recording their amplitude, timestamp, and phase angle relative to the voltage cycle's zero-crossing reference. Each pulse was assigned to a phase bin, followed by statistical aggregation of the pulse counts and amplitude distributions within each bin. This phase–amplitude–density correlation enabled the non-invasive diagnosis of insulation degradation by linking the stochastic optical behavior to the voltage waveform's periodic characteristics.

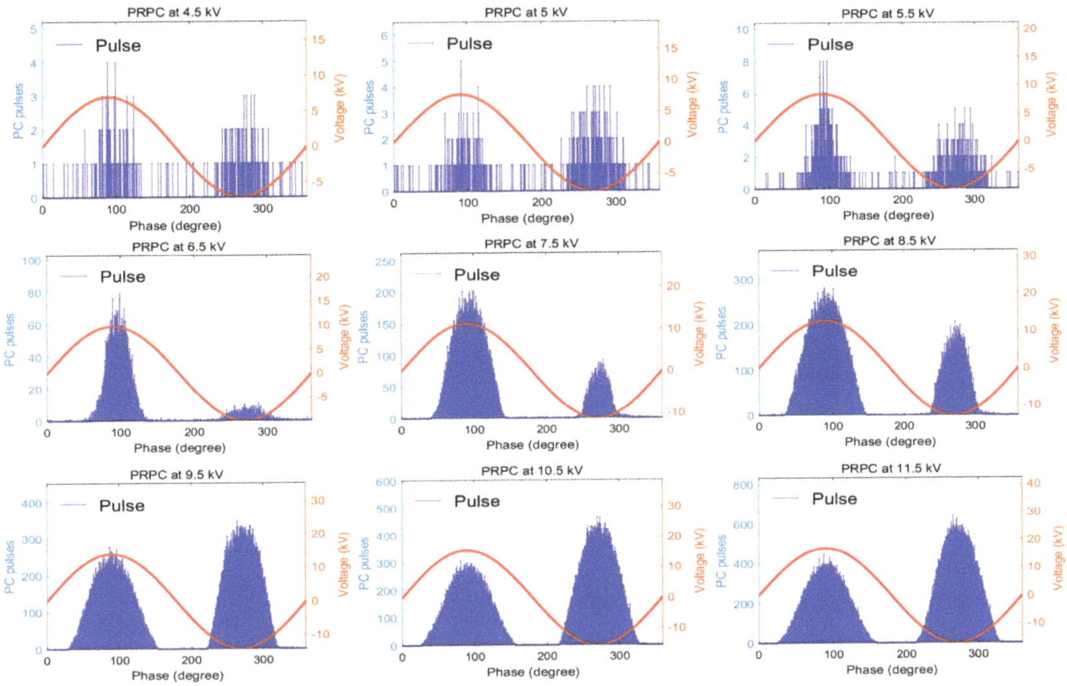

Figure 3. The PRPC of metal protrusion defects under an increasing applied voltage.

In Figure 3, each picture contains 2000 sampling points, and the photon pulse was counted in the corresponding gate time, where the unit is the photon pulse intensity (photons/microsecond). The PRPC was used to establish the polarity effect and amplitude distribution of defect-induced photon signals. As shown in Figure 3, the initial photon pulse intensity was 5 under the 4.5 kV AC voltage, and a value of merely 10 was reached when the voltage increased to 5.5 kV. However, a tenfold increase phenomenon was observed when the applied voltage reached 6.5 kV. Subsequently, the voltage increased through 12 stages, resulting in an eightfold increase in photon emission, with an average growth rate of 66%, markedly lower than the preceding phase. This can be attributed to the fact that at lower voltages, PD had not yet been initiated, and the photons originated from air ionization near the defects. Beyond 5.5 kV, PD initiation occurred, rapidly intensifying the photon emission. Further voltage escalation steadily promoted the photon emission, resulting in a smooth linear increase. A similar phenomenon was also observed in [6], in which the evolutionary process of luminescence caused by an increasing voltage was divided into four stages.

In addition, Figure 3 also illustrates the polarity effects evident in the PRPC induced by the metal protrusions. At different excitation voltages, the peak values in the spectra consistently aligned with the voltage peaks. At lower excitation voltages, the photon pulse intensity under positive polarity exceeded that under negative polarity. However, with the increase in the voltage, the PRPC revealed a steady rise in the photon pulses under a negative polarity, surpassing those under a positive polarity. This phenomenon was attributed to the following mechanism.

At lower excitation voltages, during the positive half-cycle, the protrusion with a positive charge attracted free electrons in the surrounding medium, forming a concentrated electric field region. This field was sufficient to ionize the surrounding gas molecules, generating electron–ion pairs that triggered a partial discharge. Due to enhanced electric field effects during the positive half-cycle, discharge events were more frequent, resulting in a higher photon emission. At higher voltages, both the positive and negative half-cycles sustained a sufficient electric field strength to induce a partial discharge. However, as the voltage increased, the discharge activity became more pronounced during the negative half-cycle. This was primarily due to the sustained strong electric field around the protrusion, even when it carried a negative charge during the negative half-cycle. Additionally, in the negative polarity discharge, the production and collision ionization of high-speed electrons were more efficient, resulting in increased photon emission.

3. Discharge State Analysis

3.1. Features for Characterizing the PRPC Pattern

The variation law presented in Figure 3 shows that the photons were released by different mechanisms as the voltage increased, and the distinct pattern feature caused by the PRPC can be thus used to analyze the discharge state.

It was observed that a higher applied voltage led to a higher photon pulse. Then, the photon pulse repetition rate P_{rr} is defined as follows:

$$P_{rr} = \sum_{i=1}^{N} \Phi(i)/t \tag{1}$$

where $\Phi(i)$ is the pulse intensity of sampling point i in Figure 3, t is the measurement time, and t was equal to 4 s, and N is the total number of sampling points, where $N = 2000$.

Then, to quantify the degree of dispersion of the pulse amplitude or phase distribution, the standard deviation (S_{td}) defined in Equation (2) was used.

$$S_{td} = \sqrt{\frac{\sum_{i=1}^{N} [\Phi(i) - \Phi_{aver}]^2}{N}} \tag{2}$$

Φ_{aver} is the average value of the photon pulse intensity. A high S_{td} means that there may be multiple discharge modes or unstable discharges.

In addition, skewness (S_{ke}) [18] describes the asymmetry of the distribution of PRPC, and kurtosis (K_{ur}) [19] reflects the sharpness of the distribution curve, as defined in Equations (3) and (4):

$$S_{ke} = \frac{\sum_{i=1}^{N} [\Phi(i) - \Phi_{aver}]^4}{N \cdot {S_{td}}^3} \tag{3}$$

$$K_{ur} = \frac{\sum_{i=1}^{N} [\Phi(i) - \Phi_{aver}]^4}{N \cdot {S_{td}}^4} \tag{4}$$

As depicted in Figure 3, the photon pulse in both the positive and negative AC cycles changed with the voltage, and the corresponding pulse ratio (R_{spn}) is thus defined as follows:

$$R_{spn} = \frac{\sum_{j=1}^{O} \Phi(j)}{\sum_{k=1}^{P} \Phi(k)} \tag{5}$$

In (5), O and P are the total number of pulses in the positive and negative AC cycles.

Further, to normalize the impact of the voltage, the normalized ratio of the mean photon pulse to the voltage (R_{mv}) is defined in Equation (6). U is the applied AC voltage.

$$P_{rr} = \frac{\sum_{i=1}^{N} \Phi(i)/N}{U} \tag{6}$$

In the experiment, 20 repeated measurements were taken on the defect's PRPC under the same voltage, and the corresponding data (300 samples) were then substituted into the defined formulas.

Then, the distribution between the calculated features and the voltage was plotted, as shown in Figure 4, and the fitting curves were also established to explore the variation law.

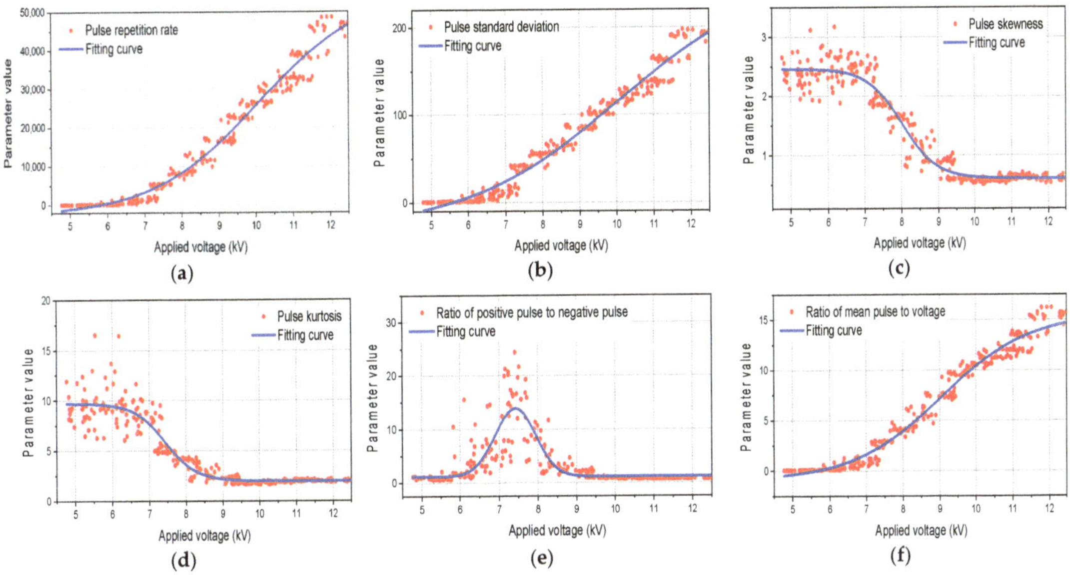

Figure 4. The distribution of the feature parameters versus the applied voltage. (**a**) Pulse repetition rate, (**b**) pulse standard deviation, (**c**) pulse skewness, (**d**) pulse kurtosis, (**e**) ratio of positive pulse to negative pulse, (**f**) ratio of mean pulse to voltage.

Figure 4 illustrates that P_{rr}, S_{td}, and R_{mv} increased gradually with the increase in the voltage. This evolution suggests that the number of photons released during the defect discharge process significantly increased as the voltage reached a certain level and gradually saturated. Conversely, S_{ke} and K_{ur} exhibited a gradual decrease and stabilization. This trend was attributed to the initial discharges from the metal protrusions potentially generating strong local discharges only at specific phase angles, hence the sharp peaks in the PRPC spectra. However, as the discharge progressed, it could occur over a wider phase range, resulting in a flatter distribution and a reduced peak height and sharpness. In addition, as shown in Figure 4e, the value of R_{spn} first increased and then decreased, consistent with the variation pattern shown in Figure 3. When the number of negative periodic photon pulses was greater than the number of positive periodic pulses, R_{spn} was less than 1.

3.2. Discharge Pattern Identification

Figures 3 and 4 demonstrate that both the spectral and quantitative features of defect-induced discharges evolved systematically with the applied voltage. Therefore, quantitative analysis based on these patterns enables the assessment of the discharge development (severity). While Figure 4 presents six distinct feature parameters, their contributions to identifying the discharge processes are rather different. To address this issue, the feature fusion method is regarded as a useful approach to solve the high-dimensional issue. t-SNE outperforms principal component analysis (PCA) [20], linear discriminant analysis (LDA) [21], and singular value decomposition (SVD) [22] (all linear methods) in nonlinear data visualization, effectively capturing the local similarities of complex manifold structures, while PCA/SVD focus on global variance and LDA relies on linear separability. Compared to UMAP [23], UMAP tends to preserve a more global structure, which may lead to blurred local details, while t-SNE's local optimization strategy makes similar clusters more compact and separates dissimilar clusters more distinctly. Additionally, UMAP is more efficient for large-scale data (greater than 10k samples), while t-SNE generates clearer cluster boundaries in smaller datasets (less than 5k samples), making it more suitable for exploratory analysis in research. Thus, a feature fusion approach based on the t-SNE model [15] was established, and the main steps include:

i. Calculate the similarity matrix: Calculate a similarity matrix based on the similarity between each pair of data points in a high-dimensional dataset.
ii. Initialization of embedding space: Randomly initialize a position for each data point in a low-dimensional space.
iii. Define t-distribution probability distribution: Use the t-distribution to define the conditional probability distribution between data points in both the high-dimensional and low-dimensional spaces.
iv. Optimization process: By minimizing the Kullback–Leibler divergence of the conditional probability distribution, adjust the position of the data in the low-dimensional space.
v. Iterative optimization: Iteratively update the position of each data point in the low-dimensional space until the stopping condition is met.
vi. Visualization and analysis: Finally, use the optimized data point positions in the low-dimensional space for visualization and analysis.

Then, the TSNE reduces the six-dimensional feature parameter matrix to three dimensions. Figure 5 illustrates the visualized distribution of features after dimensionality reduction, which indicates that the evolution stages of the discharge can be divided into three categories, which can be defined as slight discharge (SD), moderate discharge (MD), and severe discharge (SED). Then, Table 1 lists the distribution center and classification results.

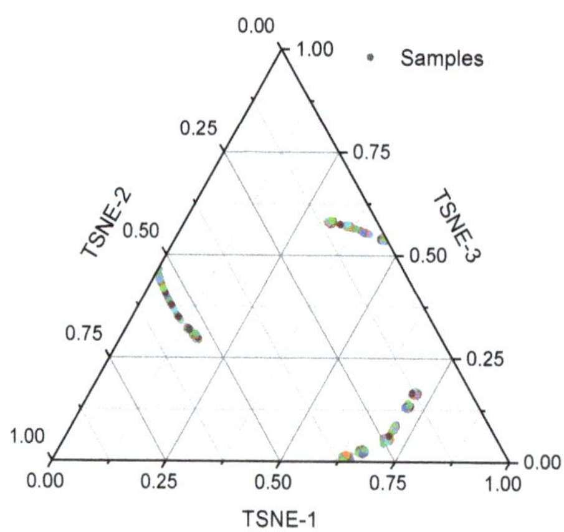

Figure 5. The sample distribution by TSNE calculation (colors indicates the diverse samples).

Table 1. The classification center and discharge state.

Item	Discharge State		
	SD	MD	SED
Classification center	(0.16, 0.59, 0.90)	(0.66, 0.96, 0.09)	(0.80, 0.10, 0.25)
Included voltages (kV)	4.5 to 6.5	7.0 to 8.5	9.0 to 11.5

By analyzing the distribution results shown in Figure 1, it was found that the samples under an excitation voltage range of 4.5~6.5 kV were located in the same region, with their corresponding coordinate center at (0.16, 0.59, 0.90). Therefore, the 100 sample points within this voltage range could be classified as the SD state. Similarly, the sample points in the second distribution region, composed of 80 samples from the 7.0~8.5 kV range, corresponded to the MD state, with their coordinate center at (0.66, 0.96, 0.09). Finally, the remaining 120 sample points belonged to the SED state, with their corresponding coordinate center at (0.80, 0.10, 0.25).

The research results show that the photon emission intensity at the defect site was positively correlated with the nearby electric field strength. A linear increase in the excitation voltage led to an exponential increase in the photon emission intensity, indicating a change in the dominant mechanism of the defect-induced luminescence process. According to previous studies [16], the iterative process of the luminescence mechanism for insulating systems should involve EL inside the solid insulation, ionization near defects, sporadic discharges near defects, and, finally, a stable discharge process. However, compared to epoxy insulating systems, the key difference in the insulating system studied here is the use of metal tip defects, meaning there is no EL process within the solid insulation. Therefore, it can be inferred that the defect-induced luminescence process in this study can be divided into three stages. The results shown in Table 1 and Figure 5 (three discharge states) confirmed this hypothesis. In summary, through clustering analysis of defect luminescence (discharge) states, the operational risks and potential hazards to GIE can be qualitatively described.

4. Conclusions

This work used photon counting to measure PD signals induced by metal protrusions and analyzed the discharge characteristics during the PD evolution process based on the PRPC pattern. The PRPC exhibited a significant polarity effect on the specimens when the applied voltage exceeded 5.5 kV, as shown in Figure 3. Specifically, the intensity of the photon pulses steadily increased with the increase in the excitation voltage, while the photons in the negative half-cycle first decreased and then increased compared to the photon pulses in the positive AC half-cycle. This phenomenon was mainly attributed to changes in AC polarity and changes in charge concentration near the needle tip. To further quantify the above changes, statistical analysis based on multiple features was proposed. Then, the quantitative relationship between the excitation voltage and these parameters was studied using fitting analysis. Finally, based on dimensionality reduction analysis, the diverse stages of the PD evolution were discussed, and their classification criteria were determined.

Given the increasing environmental requirements, research on SF_6 alternative gases has gained widespread attention. Therefore, achieving optical measurements and defect detection in different gas environments has become particularly interesting. The authors' further research will focus on the photon emission characteristics induced by defects under different gas compositions.

Author Contributions: Data curation, C.G.; Writing—original draft, H.L.; Writing—review & editing, Z.L.; Project administration, Y.L. All authors have read and agreed to the published version of the manuscript.

Funding: This work was supported by the Science & Technology Project of State Grid Corporation of China (5500-202355793A-3-8-KJ).

Data Availability Statement: Data are available on request from the authors.

Conflicts of Interest: Authors Yan Liu and Chong Guo were employed by China Electric Power Research Institute. The remaining authors declare that the research was conducted in the absence of any commercial or financial relationships that could be construed as a potential conflict of interest. The authors declare that this study received funding from State Grid Corporation of China. The funder was not involved in the study design, collection, analysis, interpretation of data, the writing of this article or the decision to submit it for publication.

References

1. Li, C.Y.; Lin, C.; Chen, G.; Tu, Y.; Zhou, Y.; Li, Q.; Zhang, B.; He, J. Field-dependent charging phenomenon of HVDC spacers based on dominant charge behaviors. *Appl. Phys. Lett.* **2019**, *114*, 202904. [CrossRef]
2. Wang, F.; Liang, F.; Zhong, L.; Chen, S.; Li, C.; Xie, Y. Short-time X-ray irradiation as a non-contact charge dissipation solution for insulators in HVDC GIS/GIL. *IEEE Trans. Dielectr. Electr. Insul.* **2021**, *28*, 704–709. [CrossRef]
3. Wang, J.; Hu, Q.; Chang, Y.; Wang, J.; Liang, R.; Tu, Y.; Li, C.; Li, Q. Metal particle contamination in gas-insulated switchgears/gas-insulated transmission lines. *CSEE J. Power Energy Syst.* **2021**, *7*, 1011–1025.
4. Albarracin, R.; Robles, G.; Martinez-Tarifa, J.M.; Ardila-Rey, J. Separation of sources in radiofrequency measurements of partial discharges using time-power ratio maps. *ISA Trans.* **2015**, *58*, 389–397. [CrossRef] [PubMed]
5. Meng, X.; Li, X.; Lu, T. Statistical properties of corona current pulses in rod-plane air gap under AC–DC composite voltages. *IEEE Trans. Dielectr. Electr. Insul.* **2024**, *3*, 212–221. [CrossRef]
6. Fan, X.; Niu, S.; Luo, H.; Liang, J.; Liu, F.; Li, W.; Liu, W.; Gao, W.; Huang, Y.; Li, C.; et al. Photon counting technique as a potential tool in micro-defect detection of epoxy insulation pull rod in gas-insulated switchgears. *High Volt.* **2024**, *9*, 267–274. [CrossRef]
7. Pandey, J.C.; Gupta, N. Study of treeing in epoxy-alumina nanocomposites using electroluminescence. *IEEE Trans. Dielectr. Electr. Insul.* **2019**, *26*, 648–654. [CrossRef]
8. Mary, D.; Albertini, M.; Laurent, C. Understanding optical emissions from electrically stressed insulating polymers: Electroluminescence in poly(ethylene terephthalate) and poly(ethylene 2,6-naphthalate) films. *J. Phys. D Appl. Phys.* **1997**, *30*, 171–184. [CrossRef]
9. Baudoin, F.; Mills, D.H.; Lewin, P.L.; Le Roy, S.; Teyssedre, G.; Laurent, C. Modeling electroluminescence in insulating polymers under AC stress: Effect of excitation waveform. *J. Phys. D Appl. Phys.* **2011**, *44*, 165402. [CrossRef]
10. Bamji, S.S.; Bulinski, A.T.; Densley, R.J. Threshold voltage of luminescence and electrical tree inception in low-density polyethylene. *J. Appl. Phys.* **2008**, *16*, 1376–1392. [CrossRef]
11. Qiao, B.; Teyssedre, G.; Laurent, C. Uncover the electroluminescence in wide band gap polymers. *J. Phys. D Appl. Phys.* **2015**, *48*, 405102. [CrossRef]
12. Yang, K.; Zhang, G.-J.; Tu, D.-M.; Yan, Z. Space charge and electroluminescence characteristics of thermally aged LDPE films. *Appl. Surf. Sci.* **2008**, *255*, 2735–2739. [CrossRef]
13. Ren, M.; Song, B.; Zhuang, T.X.; Yang, S.J. Optical partial discharge diagnostic in SF6 gas insulated system via multi-spectral detection. *ISA Trans.* **2018**, *75*, 247–257. [CrossRef] [PubMed]
14. Xia, C.J.; Ren, M.; Chen, R.F.; Yu, J.; Li, C.; Chen, Y.; Wang, K.; Wang, S.; Dong, M. Multispectral optical partial discharge detection, recognition, and assessment. *IEEE Trans. Instrum. Meas.* **2022**, *71*, 7002921. [CrossRef]
15. Fan, X.; Luo, H.; Liang, F.; Hu, J.; Liu, W.; Li, C.; He, J. Photon count technique as a potential tool for insulation micro defect detection: Principles and primary results. *iEnergy* **2023**, *2*, 258–263. [CrossRef]
16. Fan, X.; Liang, F.; Luo, H.; Zhang, X.; Huang, Z.; Li, C.; He, J. Detection of defect-induced luminescence in epoxy-based insulation materials via phase resolved photon counting method. *J. Phys. D Appl. Phys.* **2024**, *57*, 35. [CrossRef]
17. Pezzotti, N.; Lelieveldt, B.P.F.; van der Maaten, L.; Hollt, T.; Eisemann, E.; Vilanova, A. Approximated and user steerable tSNE for progressive visual analytics. *IEEE Trans. Vis. Comput. Graph.* **2017**, *23*, 1739–1752. [CrossRef]
18. Nakano, Y.; Kozako, M.; Hikita, M.; Tanaka, T.; Kobayashi, M. Estimation method of degraded vacuum in vacuum interrupter based on partial discharge measurement. *IEEE Trans. Dielectr. Electr. Insul.* **2019**, *26*, 1520–1526. [CrossRef]
19. Chen, Y.; Zhao, M.; Liu, Z.; Ma, J.; Yang, L. Comparative analysis of offshore wind resources and optimal wind speed distribution models in China and Europe. *Energies* **2025**, *18*, 1108. [CrossRef]

0. Zampolo, R.F.; Lopes, F.H.R.; de Oliveira, R.M.S.; Fernandes, M.F.; Dmitriev, V. Dimensionality Reduction and Clustering Strategies for Label Propagation in Partial Discharge Data Sets. *Energies* **2024**, *17*, 5936. [CrossRef]
1. Ji, S.; Ye, J. Generalized linear discriminant analysis: A unified framework and efficient model selection. *IEEE Trans. Neural Netw.* **2008**, *19*, 1768–1782. [PubMed]
2. Howland, P.; Park, H. Generalizing discriminant analysis using the generalized singular value decomposition. *IEEE Trans. Pattern Anal. Mach. Intell.* **2004**, *26*, 995–1006. [CrossRef] [PubMed]
3. Fan, X.; Luo, H.; Liang, F.; Ma, T.; Hu, J.; Li, C.; He, J. Feature fusion of pulse current, ultrahigh frequency, and photon count signal: A novel discharge pattern recognition method of metal particles in GIS/GIL. *IEEE Sens. J.* **2024**, *24*, 36979–36986. [CrossRef]

energies

MDPI

Article

Characterization of Space Charge Accumulations in Alternative Gas-to-Liquid Oil-Immersed Paper Insulation Under Polarity Reversal Voltage Scenarios

Ya Wang [1], Yifei Xiong [2], Zheming Wang [1] and Wu Lu [1,*]

[1] School of Electrical Engineering, Shanghai University of Electric Power, Shanghai 200090, China; wangya@mail.shiep.edu.cn (Y.W.); wangzheming1137@163.com (Z.W.)
[2] State Grid Shanghai Electric Power Company Cable Branch, Shanghai 200072, China; happybear526@163.com
* Correspondence: wuluee@shiep.edu.cn

Abstract: Due to its advantages, such as its corrosive sulfur-free property and high purity, gas-to-liquid (GTL) oil is regarded as an excellent alternative to conventional naphthenic mineral oil in the oil/paper composite insulation of UHV converter transformers. In such application scenarios, under the condition of voltage polarity reversal, charge accumulation is likely to occur along the liquid/solid interface, which leads to the distortion of the electric field, consequently reducing the breakdown voltage of the insulating material, and leading to flashover in the worst case. Therefore, understanding such space charge characteristics under polarity-reversed voltage is key for the insulation optimization of GTL oil-filled converter transformers. In this paper, a typical GTL oil is taken as the research object with naphthenic oil as the benchmark. Electroacoustic pulse measurement technology is used to study the space charge accumulation characteristics and electric field distribution of different oil-impregnated paper insulations under polarity-reversed conditions. The experimental results show that under positive–negative–positive polarity reversal voltage, the gas-impregnated pressboard exhibits significantly higher rates of space charge density variation and electric field distortion compared with mineral oil-impregnated paper. In stage B, the dissipation rate of negative charges at the grounded electrode in GTL oil-impregnated paper is 140% faster than that in mineral oil-impregnated paper. In stage C, the electric field distortion rate near the electrode of GTL oil-impregnated paper reaches 54.15%. Finally, based on the bipolar charge transport model, the microscopic processes responsible for the differences in two types of oil-immersed papers are discussed.

Keywords: voltage polarity reversal; gas-to-liquid oil; space charge; electroacoustic pulse measurement technology; bipolar charge transport model

1. Introduction

High-voltage direct current (HVDC) transmission is a key technology to realize the west-to-east transmission of electricity under the dual-carbon strategy in China, and the converter transformer, as the core equipment for power transmission in the extra-high voltage system, mainly adopts oil–paper insulation structure [1]. In this application context, traditional naphthenic oil has many disadvantages, e.g., high corrosive sulfur content, large aromatic hydrocarbon proportion, and low fire resistance [2]. Therefore, the search for more environmentally friendly and stable insulating liquids has been a core issue that requires in-depth research for converter transformers in HVDC transmission scenarios. Gas-to-liquid (GTL) oil, derived from natural gas via Fischer–Tropsch conversion, has advantages such as low aromatic hydrocarbons and high purity, and has the potential to become an alternative to naphthenic oil in converter transformers. In practical operation, converter transformers would be subjected to direct current (DC), alternating current (AC), polarity-reversed, and other complex operating voltages, of which, polarity reversal voltage is likely to cause the space charge accumulations in insulation materials, consequently distorting the local electric field and causing flashover [3]. Therefore, understanding such space charge characteristics under polarity-reversed voltage is key for the insulation optimization of GTL oil-filled converter transformers.

For the space charge measurements in oil–paper insulation, the effects of variables, such as externally applied electric field, ambient temperature, and aging degree of the testing samples on space charge and electric field characteristics inside insulation materials, have been carefully studied in the previous literature. Katsumi [4] et al. used Kerr's photoelectric technique to measure the electric field distribution of the oil–paper insulation, carried out an experimental study on 3 mm oil–paper insulation under 10 kV polarity reversal voltage, and found that the discharge was different under two polarity reversal cases, i.e., negative–positive and positive–negative voltage reversals, due to the different polarities of space charge accumulated on the surface of the oil–paper insulation. Zhang et al. measured the space charge distribution inside the oil–paper insulation at different temperatures using the pulsed electroacoustic (PEA) technique and found that the double-layer transient electric field conforms to be capacitive at room temperature, whereas the rapid change in charge density at the oil–paper interface at 60 °C leads to the double-layer transient electric field distribution to conform to be resistive [5]. Zhou et al. studied the effect of aged oil and paper on the space charge characteristics at the oil/paper interface through the PEA measurement technique, and found that, under the polarity reversal electric field, the aging state of the oil dominates the surface charge accumulation processes [6,7]. However, up to now, research on the differences between GTL oil and traditional naphthenic oils in terms of space charge accumulation and dissipation characteristics during polarity reversal is rare.

With the advances of computing technology, finite element simulation provides an effective supplement to the experimental work on space charge accumulations. He et al. used bipolar charge transport and hydrodynamic drift diffusion theory to establish a composite insulation model, and studied the impact of DC voltage amplitude, charge mobility, and other factors on spaces discharges at the oil/paper interface [8]. Based on the bipolar charge transport theory, Li et al. establishes a numerical model for calculating charge distributions along multi-layer liquid–solid insulation interfaces under AC and DC composed voltage stresses, as well as the influence of insulation board thickness and aging state on space charge accumulation rates [9].

Although a large amount of research has been conducted on the space charge characteristics of oil–paper insulation under polarity-reversed voltages, the space charge accumulation of GTL oil under voltage polarity reversal conditions and the corresponding mechanism is still unclear. In order to solve these problems, this paper builds a PEA testing system to measure the space charge and electric field distributions of GTL oil-impregnated paper under voltage polar reversal conditions, with a traditional naphthenic oil as the benchmark. The differences in space charge density, charge change rate, and electric field at the ground electrode are discussed. In addition, based on the bipolar charge transport model, a numerical model is built to calculate the differences in the charge mobilities of these two oils on space charge and electric field distributions.

2. Experimental Methodology

The insulating oil samples used in this paper are isoparaffinic GTL oil (Shell Diala S4 ZX-I) and naphthenic oil (Shell Diala S3 ZX-I), and the insulating paper is a special paperboard for transformers that meets the requirements of IEC 60554-3-5-2020 [10]. The key parameters for the two oil samples in new conditions are shown in Table 1 [11,12].

Table 1. Key parameters for GTL oil and naphthenic oil.

Items	Experimental Methods	Shell Diala S3 ZX-IG	Shell Diala S4 ZX-I
Density/(kg/m^3)	IEC 60867 [13]	878	805
Flash point/(°C)	ISO 2719 [14]	140	191
Pour point/(°C)	ISO 3016 [15]	−60	−42
Conductivity/(S/m)	IEC 60247 [16]	3.9×10^{-13}	8.9×10^{-13}
Dielectric loss factor	IEC 60247 [16]	0.00281	0.00053
Relative permittivity	IEC 60247 [16]	2.02	2.00
AC Breakdown voltage pretreatment/kV	IEC 60156 [17]	>30	70
AC Breakdown voltage after treatment/kV	IEC 60156 [17]	>70	78

Before the experiments, two types of insulating oils and paper cardboard were pretreated as follows. The insulating oils were heated at 85 °C by vacuum for 48 h to ensure the water content was less than 10 ppm.

The insulating cardboards were firstly cut into 60 × 60 × 0.5 mm slides with a total of 20. These small samples were then subjected to a drying treatment under 105 °C and 133 Pa with a duration of not less than 48 h, and then cooled down to room temperature and impregnated with dried oil samples in vacuum condition to obtain the desired oil-impregnated paper samples.

Two types of oil-immersed paper specimens were then applied with polarity reversal voltage as shown in Figure 1. The samples were first applied with +10 kV voltage for 1200 s. Then, the voltage polarity was reversed to −10 kV within 10 s and held for 1200 s. A second reversal within another 10 s restored the initial polarity voltage for 1200 s. In other words, the whole voltage polarity reversal test was divided into three stages, i.e., stage A (0–1200 s) for initial DC polarization; stage B (1200–2410 s) for the first polarity reversal to the end of negative polarity polarization; and stage C (2410–3620 s) for the second polarity reversal to the end of positive polarity polarization. The voltage reversal time met the voltage reversal time (<2 min) specified in the IEC 61378-2 standard [18] for the converter factory polarity reversal test [19]. The practical test platform is shown in Figure 2. The test temperature was controlled at 20 ± 0.1 °C by oil circulating bath, the polarity reversal voltage was applied to the semiconductor (SC) high-voltage electrode, where the specimen generated charge accumulation under the action of the applied electric field, and the grounded aluminum (Al) electrode's PVDF thin film piezoelectric transducer received the vibration acoustic wave and converted the vibration acoustic wave into the electrical signal that is linear with the distribution of the space charge [5]. The test pulse width was 10 ns and the frequency was 1 kHz. Due to the inherent attenuation characteristics of the piezoelectric sensor for receiving sound signals, the experimental results focus on the charge data at the proximal end of the grounded electrode [20].

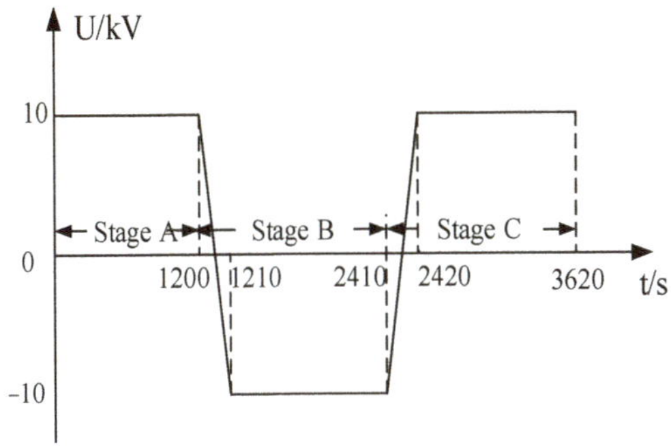

Figure 1. Polarity reversed voltage waveform.

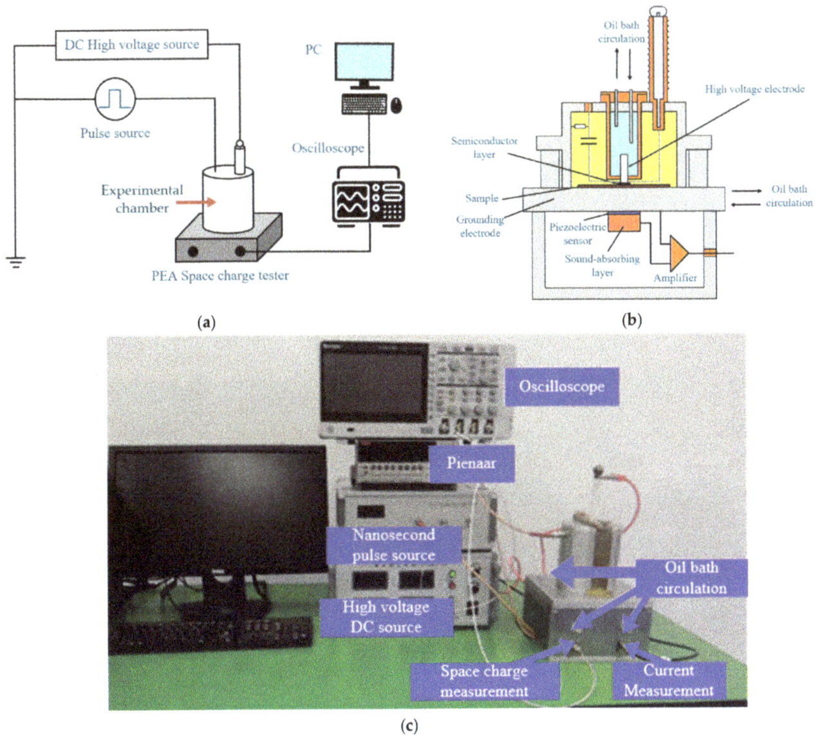

Figure 2. Space charge measuring technology. (**a**) The principal diagram for PEA testing circuit (**b**) The principal diagram for PEA testing chamber (**c**) Practical view of the experimental setup.

3. Experimental Results and Discussions

3.1. Stage A

The positive DC voltage, as shown in stage A of Figure 1, was applied to the high voltage electrode of the PEA test chamber. The internal charge distribution of the two specimens is shown in Figure 3. The space charge induced in the grounding electrode in the naphthenic oil-impregnated paper specimen decreases from $-21.08\ \mathrm{C/m^3}$ to $-18.71\ \mathrm{C/m^3}$, and the charge of the gas oil-impregnated paper specimen decreases from $-25.27\ \mathrm{C/m^3}$ to $-20.73\ \mathrm{C/m^3}$. The negative polar charge accumulated at the grounding electrode decreases gradually under the DC stresses.

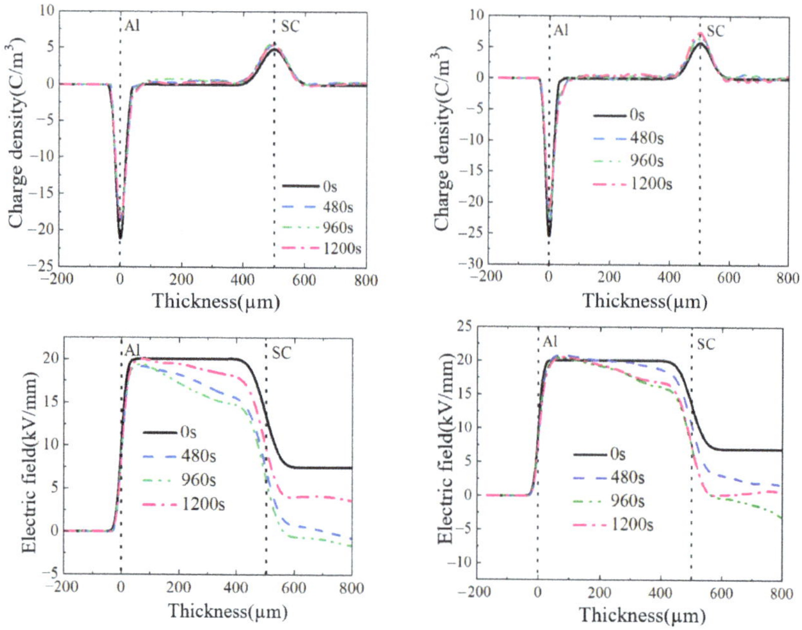

(a) (b)

Figure 3. Space charge distribution in oil-paper insulation under positive polarity voltage. (**a**) Naphthenic oil. (**b**) GTL oil.

At the initial energization time of 0 s, the space charge has not accumulated yet, the space electric field is equal to the applied electric field, with the increase in polarization time, and the uneven distribution of space charge accumulation leads to a space field strength more than 20 kV/mm. The field strength at the vicinity of the grounding electrode of the gas oil-impregnated paper is higher than that of the naphthenic oil-impregnated paper after the two specimens are polarized for 1200 s.

3.2. Stage B

Figure 4 shows the space charge accumulations of the two specimens in stage B. When the voltage polarity has not been reversed, the negative polar charge near the grounding electrode is dissipated more due to migration and recombination. At the end of the voltage polarity reversal, the space charge of the two samples at the grounding electrode reaches the maximum value, and the space charges at the grounding electrode of the naphthenic oil and GTL oil-impregnated papers are 21.9 C/m^3 and 29.4 C/m^3, respectively. The negative polar stagnant charge occurs near the grounding electrode, and the phenomenon is due to the hysteresis effect of the charge in the process of polarity reversal. The peak value of the density of the stagnant charge of the naphthenic oil- and the GTL oil-impregnated papers are −4.21 C/m^3 and −4.16 C/m^3, respectively. After 1200 s of negative DC field polarization, the space charge at the grounding electrode of GIL oil-immersed paper was reduced to 21.01 C/m^3, and that at the grounding electrode of naphthenic oil-immersed paper was 18.41 C/m^3. The field strengths of the two types of oil-impregnated paper samples are shown in Figure 4. The aberrant electric field of mineral oil reached a relatively smooth state after 1200 s of depolarization of the negative DC field, and has a high symmetry compared with the positive DC polarization for 1200 s. In contrast, the field intensity inside the specimen is lower than that before polarization after the GTL oil-impregnated paper is polarized under a negative DC field for 1200 s, whereas it does not have symmetry compared with the electric field after polarization for 1200 s under a positive polarity DC field.

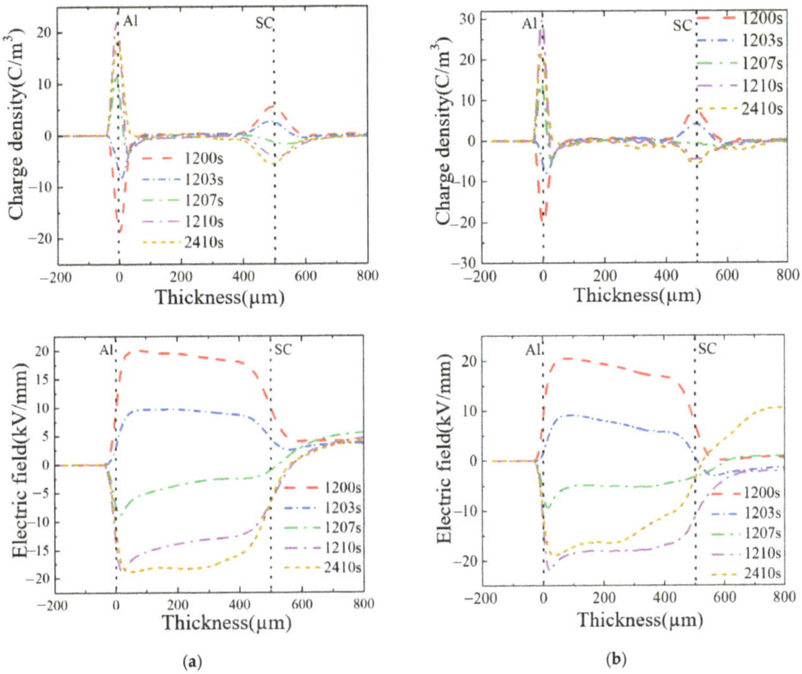

Figure 4. Space charge distribution and field strength of oil-impregnated paper under positive to negative voltage. (**a**) Naphthenic oil. (**b**) GTL oil.

3.3. Stage C

Figure 5 shows the space charge and space electric field changes in the specimens at stage C (2410–3620 s). At the end of the polarity reversal, the space charge density at the grounding electrode of naphthenic oil- and the GTL oil-impregnated papers reaches -22.2 C/m^3 and -28.42 C/m^3, respectively, whereas the stagnant positive polar charge densities at the electrodes reach the highest values of 1.7 C/m^3 and 0.6 C/m^3, respectively, with respect to the stagnant charge reductions by 2.51 C/m^3 and 3.56 C/m^3 at the polarity reversal in stage B, respectively.

After 1200 s of polarization by the positive polarity DC field, the charges at the grounding electrode of the naphthenic oil- and GTL oil-impregnated paper were reduced by -2.84 C/m^3 and -4.59 C/m^3, respectively. At the end of the applied voltage polarity reversal, there are still obvious electric field protrusions near the ground electrode in both oil–paper insulations. The peak electric field strength at the protrusion in mineral oil–paper reaches 20.45 kV/mm, whereas that in gas-to-liquid oil-impregnated paper reaches up to 23.52 kV/mm. At 3620 s, the maximum values near the grounding pole of the two specimens were 20.5 kV/mm and 26.11 kV/mm, respectively, and the field strength distortion rate in the gas-fed oil-immersed paper insulation reached 54.15%, which demonstrated easier insulation failure.

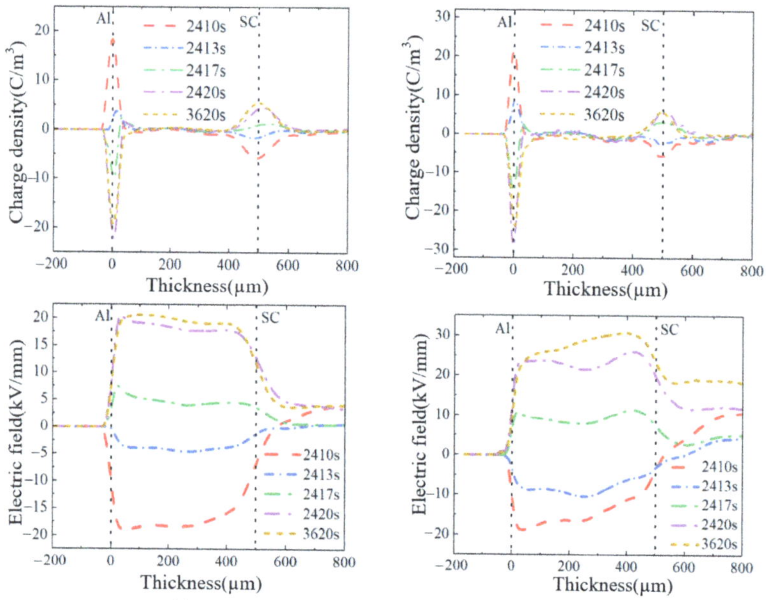

(a) (b)

Figure 5. Space charge distribution and field strength of oil-impregnated paper under negative to positive polar voltage shifting period. (**a**) Naphthenic oil. (**b**) GTL oil.

The differences in the variation in space charge at the grounding electrode between mineral insulating oil- and GTL oil-impregnated paper are shown in Figure 6, where there is a significant difference in the dynamic characteristics of the charge of the two oil-impregnated paper insulating specimens at the grounding electrode. In terms of the total amount of space charge density, space charge is injected more into the GTL oil-immersed paper. It can be seen by comparing the slope characteristics of the curves, that the rate of charge change in the naphthenic oil-impregnated paper specimens under the three electric field conditions (stage A, B, and C) is always lower than that of the GTL oil-impregnated paper specimens. The rate of space charge transport within the GTL oil-immersed paper is faster, especially in the negative polarity DC field (1210–2410 s), the charge decay characteristics of the space charge inside the gas oil-impregnated paper specimen are more obvious, and the rate of charge dissipation at the grounding electrode is about 140% higher than that of the conventional naphthenic oil-impregnated paper.

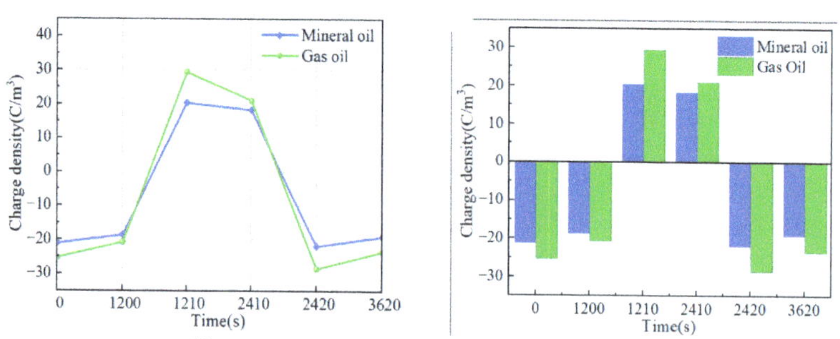

(a) (b)

Figure 6. Variation in space charge at the grounding electrode of two oil-impregnated papers. (**a**) Line graph. (**b**) Histogram.

4. Numerical Calculations for Microscopic Analyses

In the experiment, it was found that the charge dynamic properties of the two oil-impregnated papers near the grounded electrode showed obvious differences, i.e., after the first voltage inversion, the charge dissipation rate and the amount of retained charge of the two have the largest difference. This suggests that the difference in charge transport ability within the material changes the equilibrium state of charge movement and complexity and affects the charge distribution. To verify this conjecture, a bipolar charge transport model is constructed using COMSOL 6.2, a multi-physics field coupling simulation platform, to study the dynamic process of different charge carriers' mobilities during the electric field polarity reversal [21]. The right-hand side of Equations (1)–(4) describe the inter-charge carrier transitions among four morphology charge carriers, namely, free electrons ($e\mu$), trapped electrons (et), free positive charge ($h\mu$), and trapped positive charge (ht), as well as the effect on the space charge distribution, respectively. It is verified that the model can effectively reflect the characteristics of the charge behavior inside the solid medium:

$$\frac{\partial \rho_{e\mu}}{\partial t} - \nabla \cdot (\rho_{e\mu}\mu_{e\mu}E) = -S_1\rho_{ht}\rho_{e\mu} - S_3\rho_{h\mu}\rho_{e\mu} - B_e\rho_{e\mu}(1 + \frac{\rho_{e\mu}}{N_{et0}}) + D_e\rho_{et} \quad (1)$$

$$\frac{\partial \rho_{h\mu}}{\partial t} + \nabla \cdot \left(\rho_{h\mu}\mu_{h\mu}E\right) = +S_2\rho_{et}\rho_{h\mu} + S_3\rho_{e\mu}\rho_{h\mu} - B_h\rho_{h\mu}(1 - \frac{\rho_{h\mu}}{N_{ht0}}) + D_h\rho_{ht} \quad (2)$$

$$\frac{\partial \rho_{et}}{\partial t} = -S_2\rho_{h\mu}\rho_{et} - S_0\rho_{ht}\rho_{et} + B_e\rho_{e\mu}(1 + \frac{\rho_{et}}{N_{et0}}) - D_e\rho_{et} \quad (3)$$

$$\frac{\partial \rho_{ht}}{\partial t} = +S_1\rho_{e\mu}\rho_{ht} + S_0\rho_{et}\rho_{ht} + B_h\rho_{h\mu}(1 - \frac{\rho_{h\mu}}{N_{ht0}}) - D_h\rho_{ht} \quad (4)$$

where S_0, S_1, S_2, and S_3 are the composite coefficients, B_e and B_h are the trapped coefficients of free electrons and free positive charges, respectively, and D_e and D_h are the de-trapped coefficients of trapped electrons and positive charges, respectively [8].

The electron and positive charge injection densities satisfy the Schottky injection equation:

$$J_e = AT^2 \exp(-\frac{qw_{ei}}{K_bT} + \frac{q}{K_bT}\sqrt{\frac{qE}{4\pi\varepsilon_0\varepsilon_r}}) \quad (5)$$

$$J_h = AT^2 \exp(-\frac{qw_{hi}}{K_bT} + \frac{q}{K_bT}\sqrt{\frac{qE}{4\pi\varepsilon_0\varepsilon_r}}) \quad (6)$$

where A is Richardson's constant, T is the temperature set by the simulation, and K_b is Boltzmann's constant. The model is based on Poisson's equation to establish the connection between the space charge and the transient space electric field in the oil–paper insulating medium:

$$-\nabla \cdot (\varepsilon_{rp}\varepsilon_0\nabla\varphi) = \rho_{e\mu} + \rho_{h\mu} + \rho_{et} + \rho_{ht} \quad (7)$$

For the simulation, the model sets the thickness of the insulating cardboard to 0.5 mm and takes the data of a truncated line in the direction of the thickness of the insulating cardboard as the object of analysis. The simulation parameters are listed in Table 2. The mobility, as a core parameter, was reasonably configured based on the experimentally measured values from the tests in References [22,23]. The other parameter settings refer to References [8,9], where the polarity reversal voltage waveform is positive polarization voltage in 0–1200 s, polarity reversal time in 1200–1210 s, depolarization time ranges from 1210 to 3010 s, and the applied polarization field strength is 20 kV/mm. The principal model is shown in Figure 7.

Table 2. Simulation parameters for numerical calculations.

Parameters	Numerical Value	Unit	Meaning
ε_0	8.854×10^{-12}	F/m	Vacuum dielectric constant
ε_r	3.7	1	Relative dielectric constant of paperboard
B_e	0.008	1/s	Entrapment factor of free electrons
B_h	0.007	1/s	Entrapment factor for free positive charge
ω_{ei}	1.18	eV	Injection barrier for free electrons
ω_{hi}	1.20	eV	Injection barrier for a free positive charge
$\mu_{e\mu}$	5×10^{-15}	$m^2/(V{\cdot}s)$	Small mobility of free electrons
$\mu_{e\mu}$	5×10^{-13}	$m^2/(V{\cdot}s)$	Large mobility of free electrons
$\mu_{h\mu}$	4×10^{-15}	$m^2/(V{\cdot}s)$	Small mobility of free positive charge
$\mu_{h\mu}$	4×10^{-13}	$m^2/(V{\cdot}s)$	Large mobility of free positive charge
S_0	0	$m^3/(s{\cdot}C)$	Composite coefficients of incoming electrons and incoming positive charges
S_1	1×10^{-5}	$m^3/(s{\cdot}C)$	Complexity factor of free electrons with incoming positive charge
S_2	1×10^{-5}	$m^3/(s{\cdot}C)$	Composite coefficient of free positive charge and incoming electrons
S_3	1×10^{-5}	$m^3/(s{\cdot}C)$	Composite coefficient of free electrons and free positive charges
A	1.2×10^6	$A/(m{\cdot}K)^2$	Richardson's constant
K_b	1.3806×10^{-23}	J/K	Boltzmann's constant

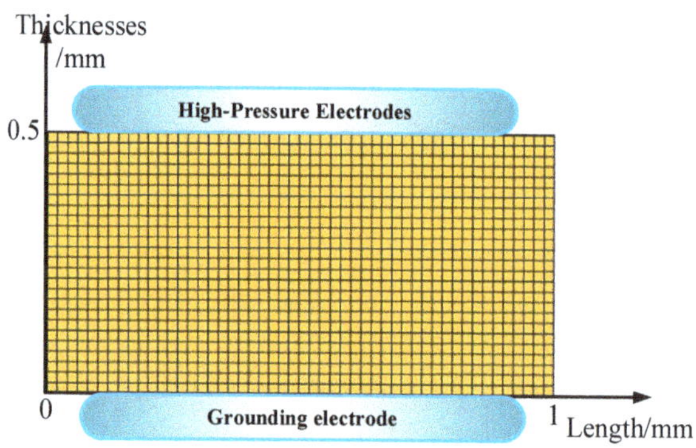

Figure 7. Oil paper compression and splitting model.

Figure 8 shows the comparison between the experimental and simulated space charges before and after polarity reversal. The correlation coefficients between the experimental and simulation data reach 0.940 and 0.794 before and after polarity reversal, respectively. These results demonstrate that the simulation exhibits good consistency with the experimental observations in characterizing the dynamic evolution trends of space charges.

It is worth noting that the spatial internal charge density obtained from the simulation is lower than the experimental value, which mainly stems from the fact that the charge trap effect existing in the actual insulating paper has not been fully modeled. The narrower range of charge distribution near the electrodes in the simulation may be related to the fact that the actual specimen has a higher trap density near the electrodes. By tuning the parameters, it is possible to isolate the influence of a single physical factor on charge transport.

Figure 9 shows the space charge distribution curves inside the insulation for different mobilities. In the initial polarization stage, the charge polarity near the electrode is dominated by the field injection mechanism and the space charge at the electrode continues to accumulate with the increase in the pressurization time. Additionally, the charge at the grounded electrode under the large and small mobility is −0.62 C/m^3 and −0.31 C/m^3, respectively, at the time of 1210 s. With the increase in free charge carrier mobility, both the spatial charge density and distribution range within the insulating material exhibit significant enhancement. This phenomenon arises because charge carriers with lower mobility are more likely to be captured by traps near the electrodes during their migration through the dielectric medium, thereby reducing internal space charge density. In contrast, rapid carrier migration promotes the deeper trapping of charges along the material thickness direction, manifesting as extended spatial charge distribution characteristics toward the dielectric bulk. Poisson's equation predicts that the maximum electric field intensity shifts to the material interior rather than remaining localized near the electrodes.

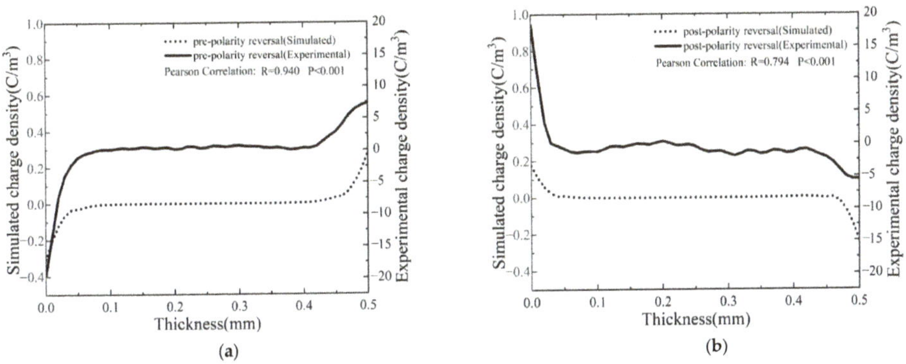

Figure 8. Comparison of experimental and simulated space charge before and after polarity reversal. (**a**) Comparison of experimental and simulated space charge before polarity reversal. (**b**) Comparison of experimental and simulated space charge after polarity reversal.

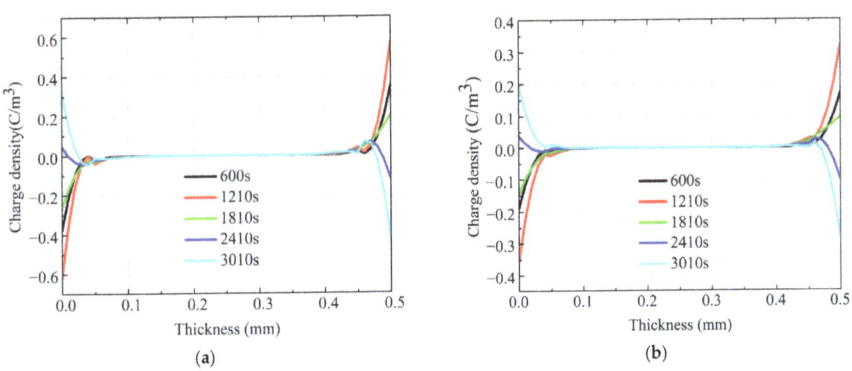

Figure 9. Spatial charge density distribution with different mobility. (**a**) Space charge density distribution at high mobility. (**b**) Space charge density distribution at small mobility.

At 1810 s, under applied voltage, polarity reversal is observed: the ground electrode exhibits positive charges while the high-voltage electrode accumulates negative charges, marking the initiation of new charge accumulation. The simulation data show that at 2410 s, the peak stagnant charge density obtained from the full neutralization of the newly injected charge with the original charge in the low-mobility material is −0.01 C/m^3. In contrast, high-mobility materials retain a higher stagnant charge density of −0.04 C/m^3 due to greater internal charge accumulation during polarization. Space charges of the same polarity concentrate near electrodes, while opposite-polarity residual charges persist in the dielectric bulk. Consequently, the interfacial field strength at the ground electrode diminishes, and the field maximum shifts toward the material interior. By 3010 s, residual charges vanish in low-mobility dielectrics, whereas high-mobility dielectrics retain pronounced charge trapping near electrodes.

Polarity reversal leads to stagnant charge density not only related to the aforementioned mobility, but also to the formation of defects in the cardboard material. The traps formed by defects are categorized into deep traps and shallow traps [24,25]. Shallow traps give rise to the short-term capture of charges and space charges are prone to de-trapping motions. Charges in deep traps require a higher energy to break free from the confinement to become free charges that can migrate, which makes the charges in space continue to pile up and trigger an electric field distortion; a distorted electric field will further interfere with the shallow trajectory of the subsequent charges. The formation of charge distribution and electric field changes shows a dynamic coupling effect. Figure 10 shows the carrier transport process inside oil-impregnated paper [26]. During DC polarization, the whole process involves a complex balance of charge injection, trapping, de-trapping, migration, and composite behavior. The stagnant charge near the electrode and the increase in the trap density is due to the polarization process. The anisotropic charge is still difficult to dissipate after the electric field is reversed and the polarity of the charge near the electrode remains unchanged for a short period of time, presenting an obvious charge stagnation phenomenon. Traps near the electrodes induce the aggregation of same-polarity charges, which partially counteract the external electric field, leading to a relative reduction in field strength near the electrode regions. Conversely, in the bulk region of the specimen, the absence of such charge compensation results in an enhanced electric field intensity, with the peak field strength typically localized within the material interior. It reflects the regulatory effect of traps in the oil–paper medium on the electric field distribution of charges.

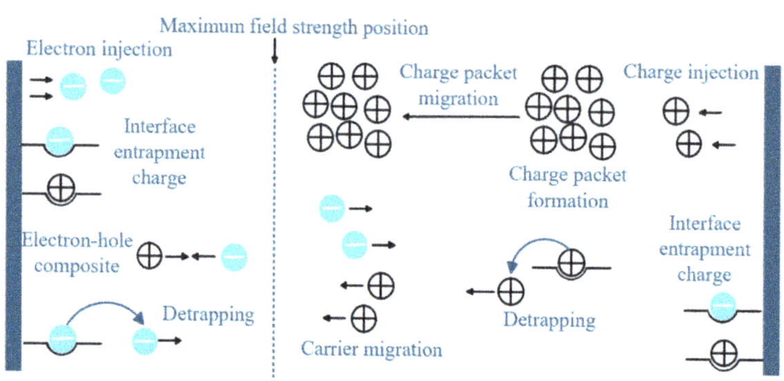

Figure 10. Charge carrier transport processes inside oil-impregnated paper.

5. Conclusions

In this paper, both experiments and simulations on the space charge accumulations in a typical GTL oil-immersed paper insulation are taken as the research object with naphthenic oil-immersed paper insulation as the benchmark. The main research findings are as follows:

(1) The experimental results demonstrate that under positive–negative–positive polarity reversal voltage, the gas-impregnated pressboard exhibits significantly higher rates of space charge density variation (140% faster negative charge dissipation at the ground electrode in stage B) and electric field distortion (54.15% distortion rate near electrodes in stage C) compared with mineral oil-impregnated paper.

(2) A simulation model based on bipolar carrier theory was developed to investigate space charge accumulation and field distortion mechanisms in oil–paper insulation

under polarity reversal. The simulation results indicate that enhanced mobility reduces space charge accumulation at the electrodes while increasing bulk charge density, thereby mitigating interfacial electric fields. However, excessive mobility may exacerbate post reversal residual charge retention, intensifying field distortion.

(3) The analysis of carrier transport mechanisms in oil-impregnated paper reveals the critical regulatory role of traps in charge/field distribution. Future research should explore modifying transformer oil–paper composites with titanium dioxide nanoparticles. This modification strategy shows potential for accelerating space charge dissipation, optimizing trap energy distribution, and alleviating interfacial field distortion in gas-impregnated pressboard [27], though its practical feasibility requires systematic investigation.

Author Contributions: Conceptualization, YW. and W.L.; methodology, Y.W.; software, Y.X.; validation, Y.W., Z.W. and W.L.; formal analysis, Y.W.; investigation, Y.X.; resources, W.L.; data curation, Y.X.; writing—original draft preparation, Y.W.; writing—review and editing, Y.W.; visualization, Y.W.; supervision, W.L.; project administration, W.L. All authors have read and agreed to the published version of the manuscript.

Funding: This research received no external funding.

Data Availability Statement: Data is contained within the article. The original contributions presented in this study are included in the article. Further inquiries can be directed to the corresponding author.

Conflicts of Interest: Author Mrs. Yifei Xiong is employed by State Grid Shanghai Cable Company Co., Ltd. The authors declare no conflict of interest.

References

1. Mao, G.L.; Zhao, F.; Li, S.J.; Ge, Z.; Li, S.T.; Yang, L.Q. Effect of Insulating Oil Types on DC Space Charge Characteristics of Oil-Paper Insulation. *High Volt. Eng.* **2022**, *48*, 1726–1736.
2. Qian, Y.H.; Huang, Y.B.; Zhang, Y.; Zhong, Z.S. Recent progress and prospects of hydrogenated isomerized non-naphthenic transformer oil. *Proc. CSEE* **2020**, *40*, 5373–5383.
3. Zhang, W.L.; Zhang, G.B. Technical specifications and key issues of transformer oil for UHV applications. *Proc. CSEE* **2009**, *29*, 1–6.
4. Kato, K.; Nara, T.; Okubo, H.; Endo, F.; Yamazaki, A.; Koide, H.; Hatta, Y.; Hikosaka, T. Space charge behavior in palm oil fatty acid ester (PFAE) by electro-optic field measurement. *IEEE Trans. Dielectr. Electr. Insul.* **2009**, *16*, 1566–1573. [CrossRef]
5. Zhang, J.F.; Dong, X.; Liu, S.L.; Yu, X.D.; Hao, L.Y.; Li, Y. Space charge characteristic of oil-paper insulation under polarity reversal voltage. *High Volt. Eng.* **2023**, *49*, 301–310.
6. Wu, Z.S.; Dai, C.; Zhou, Y.X.; Huang, M. Space charge characteristics of Nomex insulation paper and Kraft insulation paper in KI50X insulation oil. *High Volt. Eng.* **2017**, *43*, 2911–2918.
7. Zhou, Y.X.; Zhang, X.Z.; Huang, X.; Wang, H.J.; Xue, L. Influence of oil and paper aging states on space charge in oil-paper insulation. *Sci. Sin. Technol.* **2023**, *53*, 589–600. [CrossRef]
8. He, D.X.; Gong, W.J.; Liu, H.S.; Zhao, X.L.; Li, S.M.; Li, Q.Q. Simulation study on partial discharge in wedge-shaped oil gap of oil-paper insulation under DC electric field. *Proc. CSEE* **2021**, *41*, 5779–5789.
9. Li, C.; Ji, Z.; Yan, H.W. Influence of insulating paper characteristics and polarity reversal on space charge effect in oil-paper insulation. *Proc. CSEE* **2020**, *40*, 4708–4718.
10. *IEC 60554-3-5:2020*; Specification for Cellulosic Papers for Electrical Purposes—Part 3–5: Specifications for Individual Materials—Electrical Insulating Papers. IEC: Geneva, Switzerland, 2020.
11. Zhang, S.Q.; Zhao, X.L.; Qi, B.; Liu, X.; Li, C.R. Interface charge accumulation characteristics and dynamic process of paraffin-based and naphthenic transformer oil-paper under DC and polarity reversal voltages. *Trans. China Electrotech. Soc.* **2022**, *37*, 767–774, 792.
12. Wang, J.Y.; Li, J.Z.; Zhang, S.Q.; Cheng, H.C.; Gao, F.; Bai, S. Typical performance analysis of gas-to-liquid transformer oil produced from natural gas. *High Volt. Eng.* **2016**, *42*, 2974–2979.
13. *IEC 60867:2022*; Insulating Liquids—Specifications for Unused Liquids Based on Synthetic Aromatic Hydrocarbons. IEC: Geneva, Switzerland, 2022.

14. *ISO 2719:2016*; Determination of Flash Point—Pensky-Martens Closed Cup Method. ISO: Geneva, Switzerland, 2016.
15. *ISO 3016:1994*; Petroleum Products—Determination of Pour Point. ISO: Geneva, Switzerland, 1994.
16. *IEC 60247:2004*; Insulating Liquids—Measurement of Relative Permittivity, Dielectric Dissipation Factor (tan δ) and d.c. Resistivit IEC: Geneva, Switzerland, 2004.
17. *IEC 60156:2018*; Insulating Liquids—Determination of the Breakdown Voltage at Power Frequency—Test Method. IEC: Genev Switzerland, 2018.
18. *IEC 61378-2:2001*; Convertor Transformers—Part 2: Transformers for HVDC Applications. IEC: Geneva, Switzerland, 2001.
19. Gong, W.J. Charge Characteristics of Oil-Paper Insulation Under Polarity Reversal Electric Field and Its Influence on Wedg Electrode Discharge. Ph.D. Thesis, Tsinghua University, Beijing, China, 2022.
20. Ren, H.W.; Li, Q.M.; Li, C.Q.; Gao, H.Y.; Liu, T. Equivalent Transmission Line Model of Electro-Acoustic Pulse Measuremen System and Acoustic Characteristics Analysis. *High Volt. Eng.* **2021**, *47*, 1485–1495.
21. Zou, R.H.; Liao, R.J.; Hao, J.; Liu, C.; Wu, J. Simulation and deductive model of interface charge distribution in liquid-soli oil-paper insulation system under AC and AC/DC composite voltages. *High Volt. Eng.* **2021**, *47*, 1391–1402.
22. Shi, H.Z. Research on Carrier Migration Velocity in Oil-Paper Insulation under High Temperature and High Electric Fiel Master's Thesis, North China Electric Power University (Beijing), Beijing, China, 2023.
23. Gao, L.; Xu, H.; Qin, Y.; Lv, Z.; Wu, K. Influence of Composition Types of Oil-Paper Composite Insulation on Charge Transpor Characteristics. *High Volt. Eng.* **2025**, *51*, 309–321.
24. Zhang, G.Z.; Wang, K.; Yan, W.Y. Study on fiber impurity particles in liquid insulation of converter transformers. *High Volt. En* **2022**, *48*, 4297–4305.
25. Liao, R.J.; Xiang, M.; Yuan, Y.; Gao, B.L.; Zhu, T.N.; Li, W.M. Effect of Nano-Al_2O_3 doping on space charge and trap energy leve distribution characteristics of insulating paper. *High Volt. Eng.* **2019**, *45*, 681–690.
26. Gao, C.J. Research on the Generation Mechanism and Influencing Factors of Interface Charges in Oil-Paper Insulation under DC Electric Field. Master's Thesis, North China Electric Power University (Beijing), Beijing, China, 2020.
27. Hao, J.; Liao, R.J.; Chen, G.; Min, D.M.; Zhang, L.; Zou, R.H. Review of space/interface charge characteristics and suppressio methods in oil-paper composite dielectrics. *High Volt. Eng.* **2019**, *45*, 3192–3206.

energies

MDPI

Article

Cable Aging State Diagnosis Adapted from Frequency-Domain Dielectric Spectroscopy and Polarization–Depolarization Current

Yingqiang Shang [1], Yang Zhao [1], Hongquan Ji [1], Jingyue Ma [2], Jiren Chen [2,*], Ziheng Li [2], Kejia Wang [1] and Zepeng Lv [2,*]

1 State Grid Beijing Electric Power Co., Ltd., Beijing 100015, China; syq0413@outlook.com (Y.S.); 15101691629@163.com (Y.Z.); 17600036963@163.com (H.J.); 18601156800@163.com (K.W.)
2 School of Electrical Engineering, Xi'an Jiaotong University, Xi'an 710049, China; mjy19990310@stu.xjtu.edu.cn (J.M.); liziheng@stu.xjtu.edu.cn (Z.L.)
* Correspondence: cjr991224@outlook.com (J.C.); lv.zepeng.insu@xjtu.edu.cn (Z.L.)

Abstract: Cross-linked polyethylene (XLPE) cables will gradually experience aging under various stresses during long-term operation, which may lead to faults and seriously affect the safe and stable operation of the power system. This article prepares aged cable samples by accelerating the thermal aging of XLPE cables, and combines frequency-domain dielectric spectroscopy (FDS) and the polarization–depolarization current method (PDC) for detection and analysis. By measuring the dielectric loss of aged cables using frequency-domain dielectric spectroscopy, it was found that the dielectric loss value in the low-frequency region significantly increases with aging time, indicating that aging leads to an increase in polarity groups and polarization loss. The high-frequency dielectric loss also significantly increases with the strengthening of dipole polarization. At the same time, using the polarization–depolarization current method to measure the polarization current and depolarization current of cables, it was found that the stable value of polarization current increases with aging time, further verifying the changes in the conductivity and polarization characteristics of insulation materials. Combining the broadband dielectric response characteristics of FDS (0.001 Hz–1 kHz) with the time-domain charge transfer analysis of PDC, the molecular structure degradation (dipole polarization enhancement) and interface defect accumulation (space charge effect) of cable aging are revealed from both frequency- and time-domain perspectives. The experimental results show that the integral value of the low-frequency region of the frequency-domain dielectric spectrum and the stable value of the polarization depolarization current are positively correlated with the aging time, and can make use of effective indicators to evaluate the aging state of XLPE cables.

Keywords: frequency-domain dielectric spectroscopy; insulation aging; polarization depolarization current; XLPE cable

1. Introduction

Power cables are integral elements of modern power transmission and distribution systems, with their insulation status holding paramount significance for the operational reliability of the power grid. Compared with overhead lines, cable systems achieve compact urban space utilization through underground installation. However, their long-term operation under electro-thermal coupled stress leads to insulation aging, which has become a major hidden hazard threatening the safe operation of power systems [1]. Cross-linked polyethylene (XLPE), as a common insulation material, undergoes molecular chain fracture after aging, which generates free radicals and triggers chain degradation reactions. This leads to changes in crystallinity and polar groups. Such microstructural deterioration results in macroscopic property degradation through reduced mechanical performance, lowered insulation resistance, and increased dielectric loss, ultimately causing a decline in dielectric performance [2]. Statistics have shown that over 60% of cable failures are caused by insulation aging. Therefore, accurate evaluation of power cable aging has become a vital issue in power system operation and maintenance.

Traditional AC withstand voltage tests often adapt 2–3 times the rated voltage to verify insulation strength. Although the AC test can determine overall insulation withstand capability, the high voltage may accelerate dielectric damage, with the results only providing a binary judgment of pass or fail. They cannot quantify the degree of aging or provide predictive capability for insulation status. Against this background, both time-domain and frequency-domain dielectric response technologies have gained widespread attention due to their non-destructive testing characteristics. Peter Werelius applied high-voltage dielectric spectroscopy (HVDS) to study the aging characteristics of XLPE cables. The experiments revealed that aging induced voltage nonlinearity, which was a unique dielectric response manifested as significant dispersion in complex permittivity under increasing test voltages [3]. S. Hvidsten demonstrated that traditional dielectric loss measurements have significant limitations in detecting local defects in long cables, as intact sections may mask localized aging signals. They found that the nonlinearity in the dielectric spectrum was not affected by the cable length ratio [4]. Wang Wei systematically studied the effects of combined thermal and electrical aging on the insulation properties of XLPE cables. Their results showed that the dielectric loss peak increased by up to 300% in the 10^{-2}–10^{1} Hz frequency range, indicating that the accumulation of polar groups intensified dielectric relaxation [5]. Du Boxue carried out accelerated thermal aging experiments on XLPE cable press sheets and observed a strong consistency among physicochemical indicators such as oxidation induction time, carbonyl index, and elongation at break in assessing aging degree. Broadband dielectric impedance spectroscopy (BDIS) tests further revealed a strong correlation between the low-frequency dielectric spectrum and the aging state [6]. The polarization–depolarization current (PDC) method is an efficient offline and non-destructive technique for insulation diagnostics. Its principle involves applying a DC voltage to the test cable to induce polarization, followed by monitoring the variation in polarization and depolarization currents to evaluate the insulation performance and aging status. Bhumiwat developed a field-applicable, non-destructive dielectric response diagnostic method for motor insulation based on the PDC technique [7]. The study found that the unaged insulation exhibited power-law decay in PDC curves, whereas in aged insulation, polarization phenomena (e.g., interfacial water adsorption and thermal degradation byproducts) increased polarization current, deviating from the power-law behavior and exhibiting a steady-state component [7]. Dakka M.A. investigated the dielectric property changes in underground XLPE cable insulation during the aging process using the PDC method. Their study showed that the integral value of the low-frequency depolarization current increased with polarization time and varied significantly with aging [8]. Yang Fan et al. applied PDC measurements to XLPE cables at different aging stages, focusing on the low-frequency behavior of the depolarization current. The results demonstrated that aged cables had faster current decay and higher steady-state currents than new ones, indicating slower charge dissipation and the presence of residual charges [9]. Cai Gang proposed a dual-parameter diagnostic method based on PDC analysis for aged cables, incorporating both DC conductivity and nonlinear coefficients. Experimental results revealed that the aged samples exhibited a two-order-of-magnitude increase in DC conductivity, accompanied by increased nonlinearity. This combined parameter approach addressed the limitations of relying solely on conductivity, which improves the accuracy and reliability of insulation diagnostics [10].

Adapted from the PDC and FDS experimental results, this paper constructs cable aging condition parameters and performs correlation analysis with aging degree using the Pearson correlation coefficient method. We also compare the sensitivity and applicable scope of the two diagnostic techniques, providing a theoretical basis for insulation condition assessment of power cables.

2. Experimental Scheme

2.1. Thermal Aging Test

The experiment takes the YJV62 single-core cable produced by Hengrui Co., Ltd. (Lianyungang, China) as the test sample. The insulation material of the cable is made of XLPE with a rated voltage of 10 kV. In the preparation stage, it is necessary to process the cable into a three-electrode system, as shown in Figure 1. Firstly, a cable segment with a length of 0.6 m was cut and stripped of the outer sheath, armor, insulation sheath, and copper shielding layer. The exposed outer semi-conducting layer was taken as the measuring electrode. Then, a wire stripping tool was employed to remove the 4 cm material of the insulation layer, and the exposed copper conductor cores were made use of at both ends as high-voltage electrodes for measurement. Afterwards, we stripped off the 3 cm semi-conducting layers at both ends of the cable, added two copper shielding rings at each end of the insulation layer, and grounded them as protective electrodes. The semi-conducting layer must be treated as above, otherwise surface leakage current may occur, seriously influencing the experimental results' accuracy [11,12].

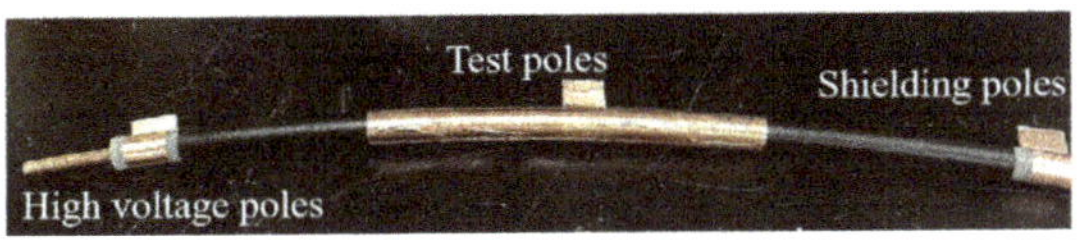

Figure 1. The cable segment sample with three electrodes used in the measurement.

The prepared cable samples were then placed in a constant-temperature aging oven for thermal aging. The acceleration temperature range is usually between 120 and 140 °C, as it is found that a higher temperature may melt the XLPE sample, while a lower temperature will not effectively accelerate the aging process. So, we chose to take 135 °C as the accelerating temperature [13]. During the test, the samples were suspended within the aging chamber to prevent contact with the chamber walls or stacking, thereby ensuring uniform exposure. For each aging period, at least three cable samples were prepared to ensure data reliability. The different sample groups were removed weekly and cooled naturally to room temperature in a dry environment before subsequent measurements were carried out.

2.2. High-Voltage Dielectric Spectroscopy Test

The IDAX-300 dielectric response analyzer produced by Megger (Fort Collins, CO, USA) was adapted in FDR measurements in this study, which allows testing in the frequency range from 10^{-4} Hz to 10^{4} Hz and maximum voltage of 2 kV. Figure 2 shows the schematic diagram of the FDS measurement. A sinusoidal test signal of 100 V was applied in the experiment, with a selected frequency sweep range of 10^{-3} Hz to 10^{3} Hz. A higher frequency cannot efficiently reflect the property change by aging, and an even lower frequency will cost too much time but provide no further information. Each measurement session lasts 60 min. During the test, the response current signal was transmitted back to the analyzer by the low-voltage electrode, which was then transferred to a computer via a data interface and processed by specialized software to derive the dielectric response characteristics of the cable insulation. The schematic diagram of the test principle is shown in Figure 3. After the test, the cable samples were grounded to discharge. Multiple repeated tests were conducted following the above steps.

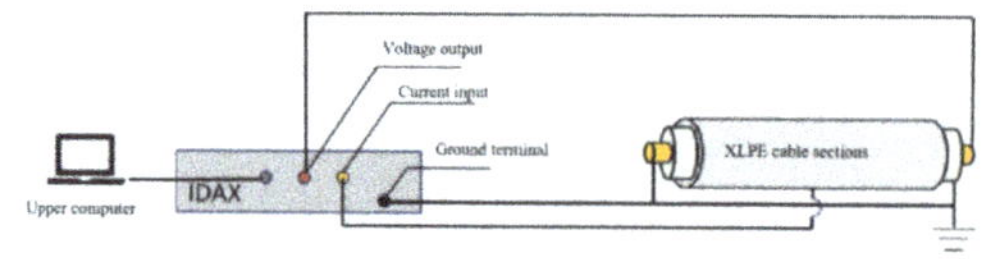

Figure 2. Schematic diagram of frequency domain dielectric spectrum testing of cable segment samples.

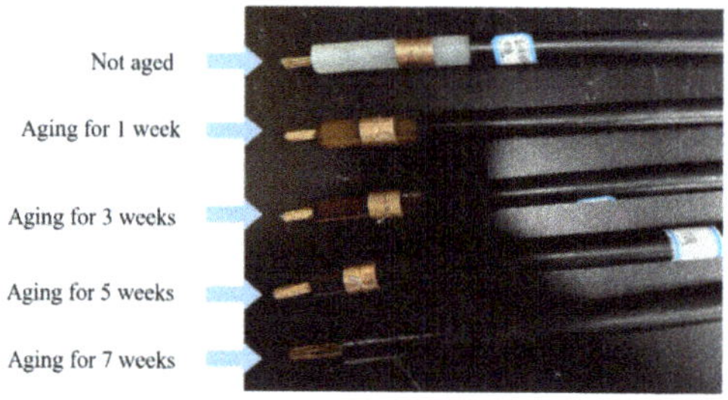

Figure 3. The cable segment sample with different thermal aging times. The cable insulation shows a darker color with longer thermal aging time.

2.3. *Polarization and Depolarization Current Test*

Figure 4 presents a schematic diagram of the PDC measurement setup. The system primarily consists of a high-voltage power supply, high-voltage relays, a protection circuit, a pico-ammeter, a shielding enclosure, a controller, and a signal acquisition and storage module.

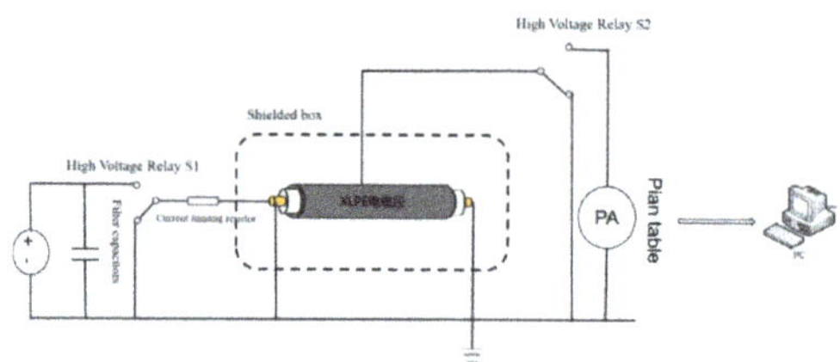

Figure 4. Schematic diagram of polarization depolarization current test for cable segment samples.

Before the experiment, the high-voltage relays S1 and S2 remained in the open state, with both the cable conductor and the outer semi-conducting layer grounded. Once the measurement started, the high-voltage DC power supply first charged the filter capacitor. Then, the relay S1 was closed to apply the polarization voltage to the cable. To prevent damage to the pico-ammeter and avoid distortion due to the initial surge current, the relay S2 was closed with a delay of 1 s after S1 closure, initiating the measurement of the polarization current. The recorded polarization current comprises the leakage current and absorption current. After the test, S1 was opened and the cable conductor was grounded, initiating the depolarization process. All current signals were transmitted via the pico-ammeter to the computer, where the polarization and depolarization waveforms were generated using the Lab-VIEW (21.0.1). A photograph of the cable PDC testing platform is shown in Figure 5.

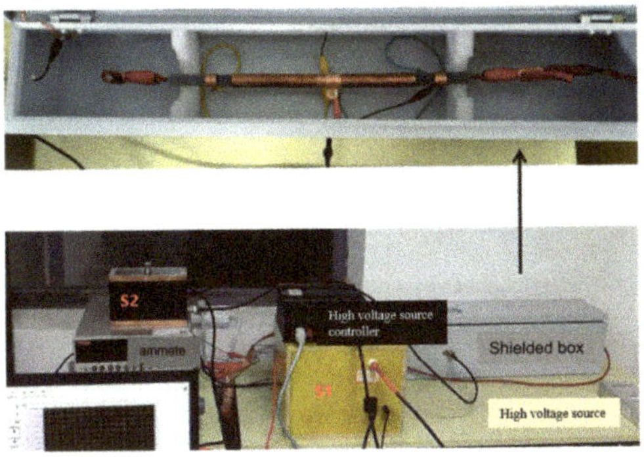

Figure 5. The photo of the polarization depolarization current testing platform.

3. Frequency-Domain Dielectric Spectrum Experimental Results and Analysis

3.1. Dielectric Spectrum Testing of Aged Cables

The dielectric spectroscopy measurement was performed on XLPE cable segment specimens subjected to different thermal aging durations by an IDAX frequency-domain dielectric response analyzer. The testing was conducted over a frequency range from 0.001 Hz to 1 kHz. To better visualize the dielectric loss factor at low frequencies, the horizontal axis is presented on a logarithmic scale. The dielectric response results of the XLPE cable segments are shown in Figure 6.

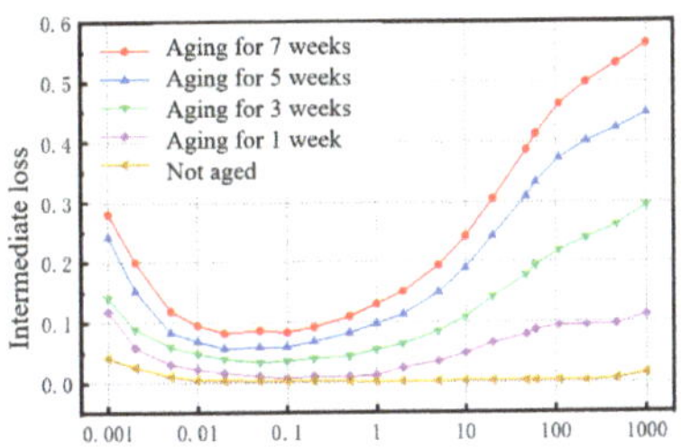

Figure 6. Frequency-domain dielectric spectrum test results of thermally aged cables. The values of dielectric loss increase significantly with the thermal aging time at 0.001 Hz and 1000 Hz.

Across the entire frequency spectrum, the dielectric loss factor (tanδ) of the unaged XLPE cable remains nearly constant, indicating a structurally uniform material with minimal frequency-dependent polarization response. As thermal aging time increases, the dielectric loss factor rises at all frequencies, which is reflected in the dielectric spectra as an overall upward shift in the curve. Simultaneously, the frequency dependence of tanδ becomes more pronounced. Particularly, the apparent increases in dielectric loss are observed at both low and high frequency extremes, especially after prolonged aging, where tanδ values become substantially higher. The growth in dielectric loss in the initial stage is relatively slow, but it accelerates obviously with the continued aging. Overall, the dielectric loss behavior of aged samples exhibits a characteristic shape in the dielectric spectrum: a relatively flat middle region with sharp increases at both ends.

In the low-frequency region, as the frequency decreases from 0.1 Hz to 0.001 Hz, the tanδ of XLPE specimens gradually increases. For unaged cables, this frequency dependence remains minimal. However, with increasing thermal aging time, the slope of the tanδ–frequency curve in the low-frequency range becomes steeper, and the difference in tanδ values among different aging durations becomes more distinct. This makes the tanδ–

f characteristics increasingly sensitive to aging, enabling more effective differentiation between varying aging states. In the most severely aged condition, the tanδ value at 0.001 Hz reaches as high as 0.37. The increase in dielectric loss at low frequency mainly corresponds to the increase in conduction current due to thermal aging. This will be discussed in the next section, together with the polarization current results.

In the high-frequency region, the dielectric loss factor (tanδ) of XLPE specimens gradually increases with rising frequency. With prolonged thermal aging, the dielectric loss becomes markedly more severe. After 7-week aging, the tanδ value peaks at 0.83 at 1000 Hz, reflecting significant deterioration in the dielectric performance of the cable insulation at higher frequencies. The increase in dielectric loss at high frequency corresponds to the increase in polar groups which enhances the reorientation polarization at relatively high frequency. The increased polar groups are considered to be generated by the broken bonds and increased C-O bond due to thermal oxidation.

3.2. Relationship Between Cable Insulation Thermal Aging Time and High-Voltage Dielectric Spectrum

An increase in polar groups in a material will increase the relative permittivity of the material. To make this clear, the relative permittivity of the cables at different aging stages was calculated at selected frequencies of 1000 Hz, 50 Hz, 1 Hz, 0.1 Hz, 0.01 Hz, and 0.001 Hz. The variation in relative permittivity with aging time at these frequencies is illustrated in Figure 7.

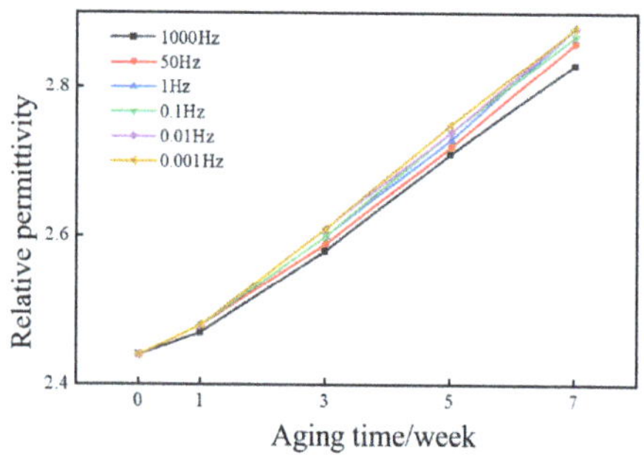

Figure 7. The changes in relative permittivity of aged cables at different frequencies. The permittivity increases with the thermal aging time, and the permittivity shows an even greater change at lower frequency.

As shown in Figure 7, the relative permittivity of XLPE cables exhibits a significant increase with prolonged thermal aging, rising from 2.44 in the unaged state to 2.99 under severe aging conditions. This change is attributed to thermal degradation, which induces molecular chain fracture and the formation of polar groups such as carbonyl and carboxylic acid groups. These polar groups possess much higher dipolar polarization capability compared to the non-polar polyethylene matrix, thereby enhancing the overall polarizability of the XLPE material and significantly increasing ε_r.

In unaged XLPE, the dielectric response is primarily governed by electronic displacement polarization and a small amount of impurity ion migration, both of which respond almost instantaneously and exhibit negligible frequency dependence over a wide frequency range. However, as aging progresses, especially by the ninth week of thermal aging, a clear frequency dependence of ε_r emerges. For severely aged cables, ε_r increases from 2.91 at 1000 Hz to 2.99 at 0.001 Hz. This frequency-dependent behavior in aged samples reflects the dynamic nature of polarization responses. The results in Figure 7 show that the relative permittivity of XLPE increases with the thermal aging. This is consistent with the increase in polar groups due to thermal aging.

To further analyze the dielectric behavior, the integral values of dielectric loss (tanδ) over the low-frequency range (0.001 Hz–1 Hz) and the high-frequency range (100 Hz–1000 Hz) were quantitatively evaluated. Figure 8 presents the trends in these integrals as a function of aging duration, highlighting the distinct evolution of dielectric loss characteristics in different frequency domains.

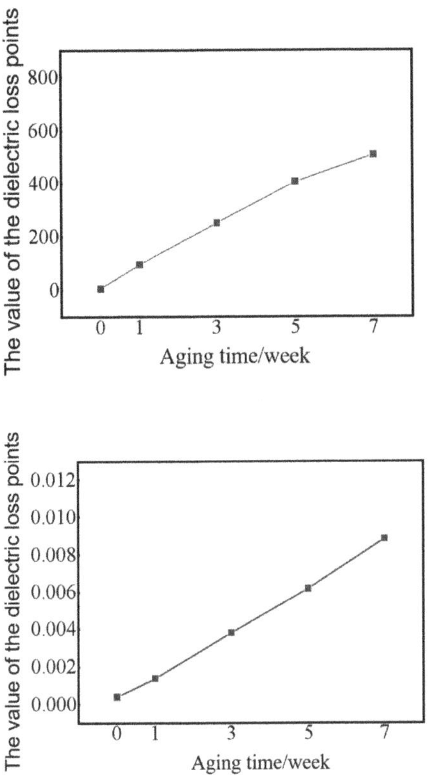

Figure 8. Integrated values of low-frequency dielectric spectra of thermally aged cables in the frequency domain. The integrated values of dielectric losses increase the changing rate of dielectric losses, and become more sensitive with the aging condition.

Comparing Figures 6–8, it can also be seen that the trend of integral values of dielectric loss show similar trends with the relative permittivity on the relationship with aging time. However, the changing rate of the integral values are much higher than the values of relative permittivity and the values of dielectric loss at a specific frequency. This shows that the integral values of dielectric loss are more sensitive with the changes in aging.

At low frequencies, dielectric loss is primarily dominated by conductive losses. As the frequency increases, the relaxation polarization cannot establish itself in time, and the dielectric material exhibits minimal relaxation loss. With increasing aging time, the dielectric loss factor at low frequencies significantly increases, with more pronounced changes occurring as aging time lengthens. Thermal aging causes the long chains of cross-linked polyethylene (XLPE) to break, generating a large number of polar groups, which leads to an increase in the low-frequency dielectric loss factor of XLPE. Furthermore, the dielectric loss increases more at lower frequencies.

In the high-frequency region, dielectric loss is mainly governed by dipolar polarization or reorientation polarization and the rapid movement of molecular chains. These processes have short response times and are highly sensitive to changes in the material's microstructure. Thermal aging results in the breakdown of the XLPE cross-linking network, leading to molecular chain fracture and making dipoles more responsive to high-frequency electric fields. The enhanced mobility of molecular chains accelerates the polarization response, resulting in an increase in high-frequency tanδ.

The dielectric loss factor shows a strong correlation with aging time, which suggests that the dielectric loss integral values can be used to characterize the overall aging state of the cable. To overcome the limitations of analyzing a single frequency range, a composite aging parameter, denoted as A2, is proposed in this study to comprehensively represent the synergistic degradation characteristics of high-frequency and low-frequency dielectric responses. The definition of this parameter is given by the following equation:

$$A_2 = \frac{1}{2} \cdot \frac{\int_{10}^{1000} \tan \delta_n(f) df}{\int_{10}^{1000} \tan \delta_0(f) df} + \frac{1}{2} \cdot \frac{\int_{0.001}^{0.1} \tan \delta_n(f) df}{\int_{0.001}^{0.1} \tan \delta_0(f) df} \quad (1)$$

Here, tanδ_{n} represents the dielectric loss value of the cable aged for n weeks, and tanδ_0 represents the dielectric loss value of the unaged cable.

Adapted from the experimental data, the criterion values for different aging periods were calculated, and the results are shown in Table 1. The criterion value A exhibits a monotonic increasing trend with aging time, rising from 9.36 to 69.28 over the 1-week to 7-week period. This demonstrates a strong correlation with the aging time. Even though the results in Table 1 are from a single experiment for each aging time, the duplicated samples show the similar values and quite the same trend.

Table 1. Integral values of dielectric loss of cable insulation with different aging times.

Aging Time	No Aging	1 Week Aging	3 Week Aging	5 Week Aging	7 Week Aging
Low-frequency integration (10^{-4})	3.8	13.9	38.1	61.8	88.6
High-frequency integration	6.457	97.28	251.96	406.25	508.87
A_2	1	9.36	24.52	39.59	51.06

The prediction error of the model is less than ±5%, indicating that the criterion A_2 effectively quantifies the aging degree of the cable insulation. Compared to the single-frequency model, the composite criterion significantly enhances the stability of the evaluation results by integrating information from multiple frequency bands. This makes it particularly suitable for complex scenarios involving non-uniform aging or local defects.

4. Polarization and Depolarization Current Testing

4.1. Polarization and Depolarization Current Tests on Cables

The polarization–depolarization current method is a commonly used technique for assessing the aging state of cable insulation. Under the same DC voltage, the longer the polarization testing time, the more thoroughly the cable insulation material becomes polarized. After applying the DC voltage, the polarization current initially decreases exponentially and then gradually stabilizes. After 20 min of voltage application, the polarization current has essentially stabilized, and the depolarization current also becomes stable after approximately 10 min. The polarization and depolarization currents at 20 min and 40 min are similar, with charge accumulation slowing down after 20 min. Considering the accuracy and convenience of the measurement, a polarization time of 20 min was selected for subsequent cable polarization–depolarization current measurement tests.

To investigate the effect of aging time on the polarization–depolarization characteristics of cable insulation materials, this study conducted current response tests on thermally aged cable samples under multiple voltage levels; 1 kV, 3 kV, and 5 kV DC voltages were applied to XLPE cable samples at different aging periods. Both the polarization and depolarization measurements lasted 20 min. The test results are shown in Figure 9.

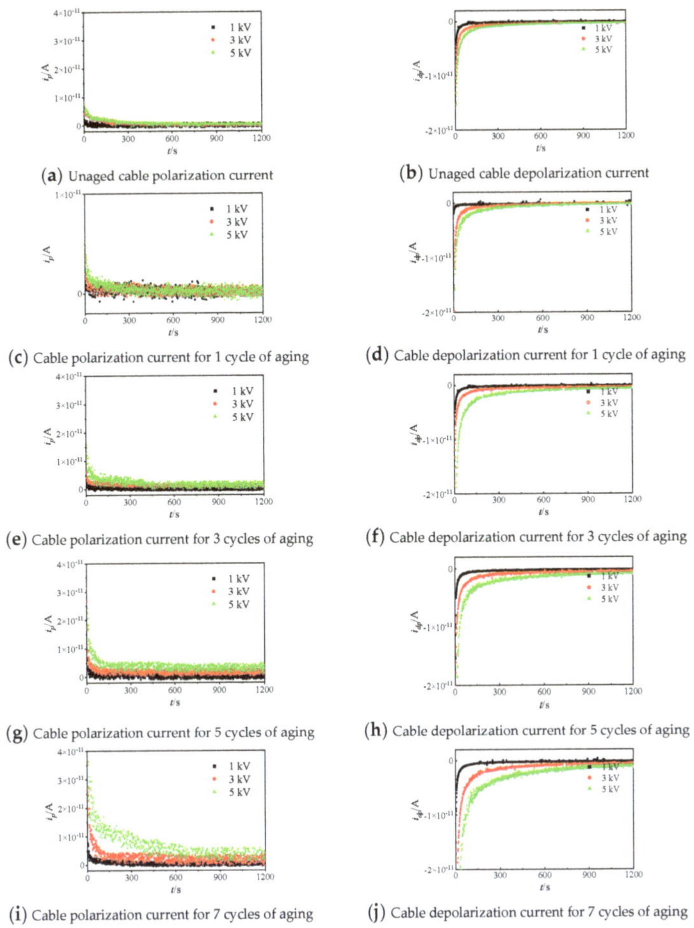

Figure 9. Polarization-depolarization current measurement results of aged cable samples under different polarization voltages.

As the voltage increases, both the polarization and depolarization current decay rates decrease, and the stable values of the polarization current increase. In the early stages of aging, the cable insulation material exhibits good performance, resulting in relatively small levels of polarization and depolarization currents. Consequently, the stable polarization–depolarization current values for the 1-week-aged cable samples are close to the noise level at all three voltage levels. Despite implementing shielding measures, external environmental and measurement equipment noise still interfered with the actual measurements. This interference is particularly noticeable in the polarization current, while the depolarization current is less affected. In the later stages of aging, significant changes in the polarization and depolarization current characteristics are observed. The polarization current's decay rate slows significantly, and at 5 kV, it takes around 15 min for the polarization current to stabilize, indicating that the dramatic increase in deep trap density is hindering charge migration. The steady-state value increases, and the difference in steady-state polarization current values becomes more pronounced with higher testing voltages, with a clear discrepancy between high and low voltage steady-state values.

The polarization current is mainly made up of the capacitance current and conduction current. The capacitance current is determined by the capacitance of the cable and the applied voltage. So, the higher the voltage, the higher the polarization current at the beginning of the voltage application. The conduction current takes the dominant part of the polarization current after several tens of seconds, once the capacitance charge on the electrodes of the cable has been fully charged. The thermal aging process can influence both the capacitance current via permittivity and the conduction current.

4.2. Relationship Between Thermal Aging Development and Polarization–Depolarization Current

To determine the influence of aging on the polarization–depolarization currents, Figure 10 compares the PDC test results for cable samples subjected to 5 kV voltage at different aging stages. Under a constant polarization voltage, as the aging degree of the cable samples increases, the measured steady-state polarization current gradually rises. For the 1-week-aged cable samples, the depolarization current decays to the noise level around 300 s, then increases due to external interference. For the 3-week- and 5-week-aged cable samples, the depolarization current stabilizes around 800 s, while for the 7-week-aged cable samples, the depolarization current stabilizes around 1200 s.

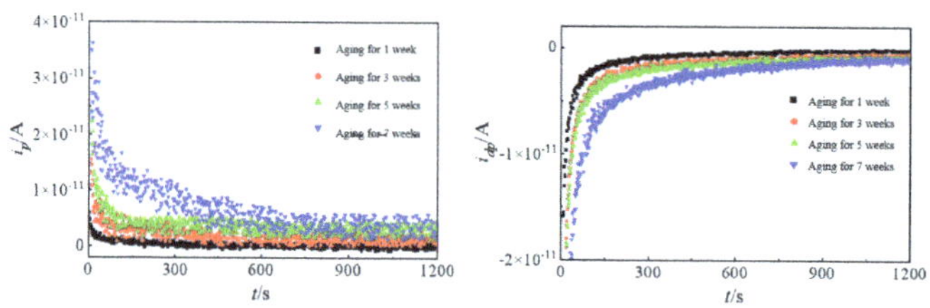

Figure 10. Polarization and depolarization currents of different aged cables under the same voltage. Both the polarization and depolarization currents increase with the aging time. The polarization current at the end of measurement time increases with the aging time, and the total dissipated charge amount during the depolarization process increases with the aging time.

As the aging time increases, the peak value of the polarization current increases, showing that the capacitance current increases with the aging time. It corresponds to the increase in relative permittivity as shown in Figure 7. The conduction current at the end of the polarization also increases with the aging time. This is due to the increase in charge density and/or mobility caused by thermal aging. The increase in conduction current also consists with the increased dielectric loss at low frequency.

The conductivity of the cable sample can be calculated using the polarization current value. The average current over the last 100 s of the polarization is taken as the conduction current. The conductivity is then calculated using the following formula:

$$\sigma = \frac{J}{E} = \frac{I}{U}\frac{ln\frac{r_2}{r_1}}{2\pi(L_1+g)} \tag{2}$$

where σ represents the conductivity ($S{\cdot}m^{-1}$), J is the conductive current density ($A{\cdot}m^{-2}$), E is the electric field strength ($V{\cdot}m^{-1}$), I is the conductive current (A), U is the applied voltage (V), r is the radius at the middle of the insulation layer (m), r_1 represents the radius of the copper core (m), r_2 is the radius at the outer surface of the insulation layer (m), L is the length of the outer semi-conductive layer (m), and g is the distance between the copper shielding ring and the outer semi-conducting layer (m).

As shown in Table 2, the calculated conductivity of the aged cable increases with the aging time.

Table 2. Conductivity of cables in different aging states.

Aging Time	1 Week	3 Weeks	5 Weeks	7 Weeks
electrical conductivity/ $\times 10^{-17}$ S·m^{-1}	1.86	7.31	8.81	11.9

The conductivity measurements at different aging durations reveal a significant increase in the electrical conductivity of aged cables. Notably, the 7-week-aged cable exhibits an order-of-magnitude enhancement in conductivity compared to the 1-week aged specimen. This phenomenon can be attributed to oxidation reactions between the insulation material and atmospheric oxygen during aging, which generate polar molecules and free radicals. These chemical byproducts act as charge carriers to enhance conductivity, thereby compromising the insulation performance of the cable. These results are also consistent with the results of lower-frequency dielectric loss. Both of the results suggest an increase in the conduction current with the thermal aging time. But the increasing rate of conductivity during aging is not the same as the low-frequency dielectric loss. This is because the conductivity obtained from the PDC is a quasi-steady state, while the lower-frequency dielectric loss obtained from FDS is a dynamic result. It is hard to directly compare the results of FDS at 0.001 Hz with the results of PDC at 1000 s.

The conduction current is determined by the charge density, mobility, and electric field. The influence of thermal aging on the conduction current of the cable insulation should relate to the change in charge density and mobility. The charge dissipates in the depolarization process are all accumulated during the polarization process. So, by integrating the depolarization current, the dissipated space charge amount during the depolarization can be obtained. It also reflects the charge accumulation inside the sample during the polarization. The formula for calculating the charge amount is as follows:

$$Q = \int_0^{t_d} \left| i_{dp}(t) \right| dt \tag{3}$$

In the formula, Q represents the total charge (C), t_d is the total time during the depolarization phase (s), and $i_{dp}(t)$ is the depolarization current (A).

The charge density of the cable refers to the charge per unit volume within the cable, which can be calculated by dividing the total charge of the cable by the volume of the cable's insulation layer. The specific formula is:

$$\rho = \frac{Q}{\pi(r_2^2 - r_1^2)L} \tag{4}$$

where ρ represents charge density (C·m^{-3}).

Currently, there is no feasible experimental method to directly measure the carrier mobility. As the conductivity and charge density have been obtained with the above method, the apparent charge mobility can be calculated as follows:

$$\mu = \frac{\sigma}{\rho} \tag{5}$$

where μ represents carrier mobility ($m^2 \cdot (V \cdot s)^{-1}$).

In the depolarization test after a 5 kV polarization, the charge amount and carrier mobility of cable samples with different aging times were calculated by integrating the depolarization current, as shown in Table 3. It was observed that both the total charge and the carrier mobility increased as the aging time increased, which is consistent with the changes in conductivity. It should be noted that the charge density and charge mobility did not change linearly with the aging time. At the beginning of the aging, the charge amount and mobility increased sharply in the first 3 weeks, and then increased slightly afterwards. The increase in charge density is mainly due to the degradation of insulation causing a higher charge injection rate at the electrode interface. The increased charge mobility is probably due to the weaker charge blocking ability with more broken bonds.

Table 3. The charge amount, density, and mobility of insulation with different aging times.

Aging Time	1 Week	3 Weeks	5 Weeks	7 Weeks
The total amount of charge/C	1.5×10^{-9}	2.6×10^{-9}	2.8×10^{-9}	3.3×10^{-9}
Charge density/$C \cdot m^{-3}$	8.74×10^{-6}	1.52×10^{-5}	1.63×10^{-5}	1.92×10^{-5}
Charge mobility/$m^2 \cdot (V \cdot s)^{-1}$	2.13×10^{-12}	4.81×10^{-12}	5.41×10^{-12}	6.19×10^{-12}

5. Correlation Analysis of Cable Aging State Parameters

To evaluate and compare the capability of different aging state parameters in characterizing aging time, statistical analysis was conducted by the Pearson correlation coefficient (r) and significance testing. The Pearson correlation coefficient was employed to quantify the linear correlation between two variables, as in:

$$r = \frac{n\sum x_i y_i - \sum x_i - \sum y_i}{\sqrt{n\sum x_i^2 - (\sum x_i)^2} \cdot \sqrt{n\sum y_i^2 - (\sum y_i)^2}} \quad (6)$$

The conductivity, charge density (ρ), carrier mobility (μ), total charge (Q), low-frequency integral, high-frequency integral, and aging parameter A of aged cable were analyzed for their correlation with aging time, with a significance level set at $\alpha = 0.05$. The analysis results are shown in Table 4.

Table 4. Correlation analysis of aging state parameters.

	Pearson Correlation Coefficient (r)	p
Electrical conductivity σ	0.972	0.0281
Charge density ρ	0.750	0.0250
Carrier mobility μ	0.803	0.0197
The total amount of charge Q	0.952	0.0479
Low-frequency integration	0.990	0.0157
High-frequency integration	0.992	0.1255
Aging parameter A	0.998	0.0438

The results indicate that all parameters exhibit strong positive correlations with aging time ($r > 0.7$), confirming their reliability as indicators for insulation aging characterization. In the FDS test, the high-frequency integral ratio shows a correlation coefficient of $r = 0.992$ with aging time, demonstrating the sensitivity of high-frequency components to microscopic mechanisms of thermal aging. The low-frequency integral ratio also shows a strong positive correlation ($r = 0.990$), although its growth rate is slower, reflecting the gradual accumulation of deep polarization effects. The aging parameter A, which integrates both high- and low-frequency features, achieves an enhanced correlation of

$r = 0.998$, outperforming single-frequency-band analyses. The evolution pattern of dielectric spectral integrals during thermal aging has clear physical significance of high-frequency integral domination in early degradation, while low-frequency integrals reflect long-term polarization damage. The proposed parameter A not only enables effective integration of multi-frequency information but also provides a theoretical foundation for rapid insulation assessment in practical application.

In the PDC test, charge amount, charge density, carrier mobility, and conductivity all increase significantly with aging time. Among these, conductivity has the highest correlation coefficient ($r = 0.972$), indicating its highest sensitivity to aging status. Not only does conductivity exhibit strong correlation, but its value also increases exponentially with aging time, making it a clear differentiator of aging stages and the most effective parameter for evaluating insulation aging status.

6. Conclusions

In this study, thermal aging of XLPE cables is investigated through accelerated aging experiments of varying durations. Samples from different aging cycles are subjected to FDS and PDC tests. The main findings are concluded as follows:

(1) FDS measurement on aged cables reveals a general upward trend in dielectric loss. The low-frequency dielectric loss factor of XLPE increases with aging duration. The lower the frequency, the greater the increment in dielectric loss. The integral of the dielectric loss from FDS measurement also shows a monotonic increase with aging time, indicating that the integral value serves as an effective indicator for XLPE insulation aging state evaluation.

(2) From the PDC tests, it can be seen that the decay rates of the polarization and depolarization currents significantly decrease after aging, while the steady-state values increase. DC conductivity, space charge density, and carrier mobility are calculated and adapted from the measurement results. Both conductivity and carrier mobility increase notably with aging duration, suggesting that these parameters can be used as reliable indicators of cable insulation aging.

(3) Correlation analysis between aging time and parameters extracted from FDS and PDC tests, including conductivity, charge density (ρ), carrier mobility (μ), charge amount (Q), low-frequency integral, high-frequency integral, and aging parameter (A) shows the strongest correlation between the conductivity and the aging time ($r = 0.9$, $p < 0.05$). Outperforming charge density and mobility, the DC conductivity is better suited for aging assessment. In FDS tests, the high-frequency integral exhibits a non-significant correlation with a significance level $p > 0.05$, whereas the low-frequency integral demonstrates a strong correlation coefficient of 0.99 ($p < 0.05$), indicating a robust relationship with cable aging degradation. This establishes the low-frequency integral as a reliable metric for characterizing the insulation aging state of cables.

Author Contributions: Conceptualization, Y.S.; Methodology, Y.S. and Z.L. (Ziheng Li); Validation, Y.Z., H.J., J.M. and J.C.; Investigation, Z.L. (Ziheng Li) and K.W.; Data curation, Y.S., J.M. and Z.L. (Ziheng Li); Writing—original draft, Y.S., J.M. and J.C.; Writing—review & editing, Z.L. (Zepeng Lv). All authors have read and agreed to the published version of the manuscript.

Funding: This research was funded by the Science and Technology Project of State Grid Beijing Electric Power Company, titled Research and Application of Cable Insulation Defect Detection and Aging Assessment Technology Based on Frequency Domain Dielectric Spectroscopy (Project Code: 520246230008).

Data Availability Statement: The original contributions presented in the study are included in the article, further inquiries can be directed to the corresponding author.

Conflicts of Interest: Authors Yingqiang Shang, Yang Zhao, Hongquan Ji and Kejia Wang were employed by the company State Grid Beijing Electric Power Co., Ltd. The remaining authors declare that the research was conducted in the absence of any commercial or financial relationships that could be construed as a potential conflict of interest.

References

1. Du, B.X.; Li, Z.; Yang, Z.; Li, J. Application and research progress of HVDC XLPE cables. *High Volt. Eng.* **2017**, *43*, 14–24.
2. Zhou, Y.; Zhao, J.; Liu, R.; Chen, Z.; Zhang, Y.X. Key technical analysis and prospect of high voltage and extra-high voltage power cable. *High Volt. Eng.* **2014**, *40*, 2593–2612.
3. Werelius, P. *Development and Application of High-Voltage Dielectric Spectroscopy for Diagnosis of Medium-Voltage XLPE Cables*; Kungliga Tekniska Hogskolan: Stockholm, Sweden, 2001.
4. Hvidsten, S.; Ildstad, E.; Sletbak, J.; Faremo, H. Understanding water treeing mechanisms in the development of diagnostic test methods. *LEEE Trans. Dielectr. Electr. Insul.* **2002**, *5*, 754–760. [CrossRef]
5. He, D.; Gu, J.; Lu, J.; Wang, W. Electrical-thermal ageing test design of 10 kV XLPE cable. *Insul. Mater.* **2016**, *49*, 53–58.
6. Xue, C. *Research on Assessment Method for Insulation Aging State of XLPE Cables*; Tianjin University: Tianjin, China, 2014.
7. Bhumiwat, S.A. On-site non-destructive dielectric response diagnosis of rotating machines. *IEEE Trans. Dielectr. Electr. Insul.* **2010**, *17*, 1453. [CrossRef]
8. Abou Dakka, M.; Bulinski, A.; Bamji, S.S. On-site diagnostics of medium-voltage underground cross-linked polyethylene cables. *IEEE Electr. Insul. Mag.* **2011**, *27*, 34–44. [CrossRef]
9. Yang, F.; Shen, Y.; Wang, Y.; Zhao, A.; Zhang, X.; Deng, J.; Zhang, G. Thermal Aging Status Diagnosis of XLPE Cable Using Polarization/Depolarization Current Method. *High Volt. Eng.* **2016**, *42*, 496–503.
10. Cai, G.; Liu, X.; Pu, J.; Zhou, K.; Huang, M. Insulation diagnosis of water-treed cables based on polarization and depolarization current method. *Insul. Mater.* **2016**, *49*, 61–64+71.
11. Lu, C.; Guangmao, L.; Sen, Y. Review of research on frequency domain response method and its application in defect location of power cables. *South. Power Syst. Technol.* **2023**, *17*, 53–64.
12. Zhou, K.; Yuan, H.; Li, Y.; Li, M.; Li, Z.; Lin, S. Assessing aging status and type of XLPE cable insulation with a graphic approach based on PDC measurement. *IEEE Trans. Power Deliv.* **2022**, *37*, 5114–5123. [CrossRef]
13. *GB/T 11026.1-2012*; Electrical Insulating Materials—Thermal Endurance—Part 1: Aging Procedures and Evaluation of Test Results. Jointly Issued by the General Administration of Quality Supervision, Inspection and Quarantine of the People' Republic of China: Beijing, China, 2012.

energies

MDPI

Article

Estimation of Remaining Insulation Lifetime of Aged XLPE Cables with Step-Stress Method Based on Physical-Driven Model

Yingqiang Shang [1], Jingjiang Qu [1], Jingshuang Wang [1], Jiren Chen [2,*], Jingyue Ma [2], Jun Xiong [1], Yue Li [2] and Zepeng Lv [2,*]

1 State Grid Beijing Electric Power Co., Ltd., Beijing 100015, China; syq0413@outlook.com (Y.S.); 15611709968@163.com (J.Q.); wjs19730613@163.com (J.W.); xiongjunncepu@163.com (J.X.)
2 School of Electrical Engineering, Xi'an Jiaotong University, Xi'an 710049, China; mjy19990310@stu.xjtu.edu.cn (J.M.); liyue1231@foxmail.com (Y.L.)
* Correspondence: cjr991224@outlook.com (J.C.); lv.zepeng.insu@xjtu.edu.cn (Z.L.)

Abstract: The remaining lifetime of the cable insulation is an important but hard topic for the industry and research groups as there are more and more cables nearing their designed life in China. However, it is hard to accurately and efficiently obtain the ageing characteristic parameters of cross-linked polyethylene (XLPE) cable insulation. This study systematically analyzes the evolution of the remaining insulation lifetime of XLPE cables under different ageing states using the step-stress method combined with the inverse power model (IPM) and a physical-driven model (Crine model). By comparing un-aged and accelerated-aged specimens, the step-stress breakdown tests were conducted to obtain the Weibull distribution characteristics of breakdown voltage and breakdown time. Experimental results demonstrate that the characteristic breakdown field strength and remaining lifetime of the specimens decrease significantly with prolonged ageing. The ageing parameter of the IPM was calculated. It is found that the ageing parameter of IPM increases with the ageing time. However, it can hardly link to the other properties or physic parameters of the material. The activation energy and electron acceleration distance of the Crine model were also calculated. It is found that ageing activation energy stays almost the same in samples with different ageing time, showing that it is a material intrinsic parameter that will not change with the ageing; the electron acceleration distance increases with the ageing time, it makes sense that the ageing process may break the molecule chain of XLPE and increase the size of the free volume. It shows that the Crine model can better fit the physic process of ageing in theory and mathematic, and the acceleration distance of the Crine model is a physical driven parameter that can greatly reflect the ageing degree of the cable insulation and be used as an indicator of the ageing states.

Keywords: ageing model; Crine model; cross-linked polyethylene (XLPE); remaining lifetime; step-stress method

1. Introduction

Due to its excellent insulation performance, mechanical strength, and thermal conductivity, cross-linked polyethylene (XLPE) has been widely used in high-voltage direct current cable insulation systems and serves as the primary insulating material for 10 kV to 220 kV power cables [1]. With the continuous growth of societal electricity demand, requirements for cable operational stability have become increasingly stringent. As early installed cables approach extended service periods, many cables deployed before 2003 have now operated for over 20 years. The failure rates of the cable will rise with the reduction in the insulation lifetime, as the bathtub curve and the application data suggest. So, the remaining ageing lifetime of these old cables has become a critical concern for the cable operators. Effective monitoring of XLPE cable insulation and accurate assessment of its remaining lifetime are essential for enhancing power system reliability.

High-voltage cables are typically designed for a 30- or 40-year service life. However, factors such as manufacturing processes, installation quality, and operational stresses (electrical, thermal, and mechanical) significantly impact XLPE insulation performance [2]. Statistical analyses of cable failures reveal that operational failure rates follow a bathtub curve. With many XLPE cables nearing their 30-year design lifetime, insulation degradation has become increasingly prominent [3]. Harsh installation environments and inherent local defects further accelerate ageing, leading to frequent insulation failures. The growing number of ageing cables in service has imposed substantial pressure on maintenance operations.

Merely evaluating XLPE insulation status proves insufficient for precise remaining lifetime prediction. Premature cable replacement incurs unnecessary economic costs, while delayed action jeopardizes system safety. Thus, developing an engineering-applicable cable lifetime assessment methodology enables scientifically optimized maintenance planning and targeted resource allocation, minimizing wasteful expenditures.

Extensive research has been conducted globally on insulation material lifetime evaluation, with models broadly categorized into empirical and physical types. Empirical models like the Dakin model and inverse power model, derived from statistical electrical ageing patterns, are widely used in industrial testing and ageing assessments due to their simplicity and ability to reflect characteristic ageing trends [4,5]. However, these models lack direct correlation with intrinsic ageing mechanisms. For instance, the ageing lifetime exponent in the inverse power model remains an empirical parameter without explicit physical meaning or connection to material properties.

To address these limitations, researchers have developed physical models based on microscopic processes, including the Lewis kinetic model [6], DMM space charge model [7], and Crine thermodynamic model [8]. The Lewis model interprets ageing as a chemical dynamic process where charge carriers gain energy under electric fields, causing structural damage through impact ionization and bond rearrangement. The DMM model emphasizes space charge effects in insulation ageing, proposing a lifetime model based on space charge dynamics. The Crine model conceptualizes ageing as a thermally activated process requiring overcoming energy barriers [8].

The Crine model frames polymer ageing as a thermally activated transition from unaged to aged states, necessitating the surmounting of ageing activation energy barrier [8]. Electric fields reduce the required activation energy by introducing electrostatic forces from charged particle. The ageing activation energy barrier exhibits linear variation with electric field intensity, influenced by the ageing activation energy difference between initial and final states. It is considered that a critical field strength exists at the exponential-to-non-exponential transition point, beyond which micro-void or defect formation may initiate. Below this threshold, ageing is suppressed, though sub-threshold ageing remains possible under specific conditions.

While the inverse power model retains broad applicability, it is an empirical model with good mathematic fitting with the experimental results but hardly links to the physics of the insulation materials. Moreover, it cannot be sufficiently used in the estimation of ageing states of the insulation.

This study compares the inverse power model and Crine model with step-up test data, investigates Crine-based ageing mechanisms, and explores methods of assessing insulation ageing states using the Crine model.

2. Experimental Setup

2.1. Sample Preparation

The cross-linked polyethylene (XLPE) insulation samples are obtained from a YJV62 single-core cable with a rated voltage of 10 kV, produced by Jinda Cable from Tianjin, China. The XLPE samples are made by a numerically controlled machine tool via ring cutting. All the samples have a thickness of 0.2 mm, as shown in Figure 1. A portion of the un-aged cable insulation layer from the middle section was ring-sectioned to obtain cable insulation slice samples. The obtained XLPE cable insulation material slices and remaining un-aged cable sections were subjected to thermal ageing in an oven. The acceleration temperature range is usually set between 120 and 140 °C, as it is found that a higher temperature may melt the XLPE sample and a lower temperature will not effectively accelerate the ageing process. So, we choose to use 135 °C as the accelerating temperature [9]. The cables were thermally aged for 0, 3, and 5 weeks. The aged cables were sectioned to obtain slices of the aged cable insulation material, which were labelled according to the ageing weeks, as AS0, AS3, AS5, etc.

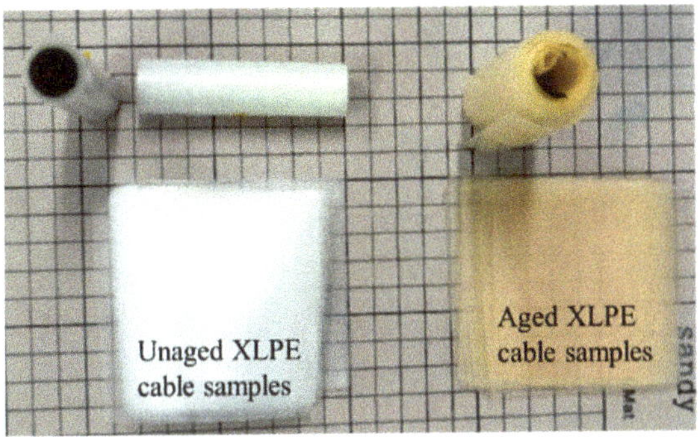

Figure 1. XLPE cable insulation slice sample.

The ring-sectioned cable slices are inherently curved, uneven, and marked with cutting scratches, which adversely affect the accuracy of experimental measurements. To obtain transparent and flat slice samples, the cable slices were placed in a 0.2 mm thick stainless-steel mould, with silicone oil-coated PET films placed on both the top and bottom surfaces to prevent adhesion. After heating the flat-plate vulcanizing press to 120 °C, the samples were placed inside and pressed under a pressure of 15 MPa for 40 s. Upon removal, they were sandwiched between water-cooled plates for cooling.

2.2. Accelerated-Ageing Test Platform

Acquisition of ageing lifetime exponents or other critical parameters for insulation material lifetime models necessitates accelerated-ageing tests, which predominantly employ the constant voltage method and step-up voltage method—each offering distinct advantages suited to different testing scenarios. The constant voltage method applies a steady voltage until specimen breakdown, providing simplicity and reliability but requiring extended test durations with inherent limitations including time inefficiency, low data acquisition rates under low-voltage conditions, and significant data dispersion. In contrast, the step-up voltage method, grounded in the cumulative ageing lifetime model, operates on the premise of irreversible ageing damage accumulation in insulation materials. This approach demonstrates superior efficiency, shorter testing cycles, and reduced data scatter. By incrementally elevating the applied voltage at predetermined time-steps until breakdown occurs, this method capitalizes on the cumulative effects inherent to solid insulation, particularly effective for materials lacking self-healing properties.

The Nelson model postulates that material residual lifetime is determined by the synergistic effects of voltage stress and breakdown probability at corresponding stress levels [10]. By conceptualizing insulation breakdown as a consequence of voltage stress accumulation, progressive voltage escalation enables evaluation of solid insulation's endurance capability and remaining service life.

Figure 2 schematically illustrates the step-up voltage ageing test configuration with the following definitions:

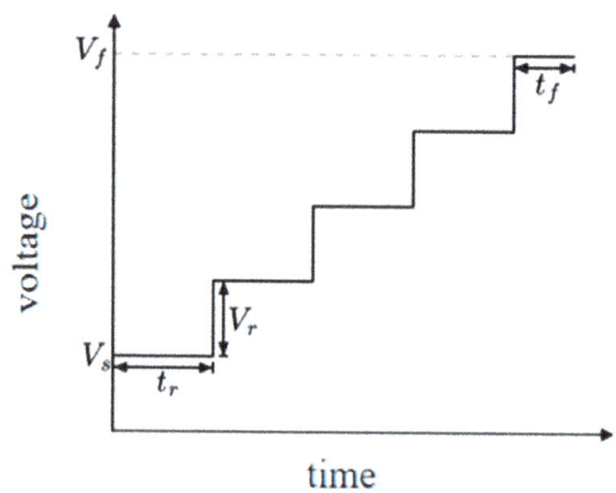

Figure 2. Schematic diagram of gradually increasing pressure method for accelerating ageing test pressure.

V_s: Initial voltage;
T_r: Time step duration;
V_r: Voltage increment per step;
V_f: Breakdown voltage;
T_f: Final step duration.

In accelerated-ageing tests the prepared sheet samples are cleaned with alcohol and placed between the upper and lower electrodes, ensuring the samples are fully immersed in insulating oil. First, a constant voltage breakdown test is conducted at room temperature to determine the sample's breakdown voltage range. For the step-stress method, the initial voltage is typically set to 60% of the room-temperature breakdown voltage [11]. The voltage gradient is selected to ensure at least three test steps before breakdown occurs, with gradient time intervals (e.g., 5 min, 10 min, 30 min) chosen. Voltage is continuously increased until the sample breaks down, and the breakdown time and voltage are recorded. The accelerated-ageing test setup is illustrated in Figure 3. In this study, the initial voltage was set to 45 kV, the voltage gradient to 5 kV, and three time-steps of 200 s, 600 s, and 1800 s were applied. The temperature is 25 °C, and the humidity is about 40%. To ensure the accuracy of the test, 6 samples are used in each group to reduce the dispersity of the data.

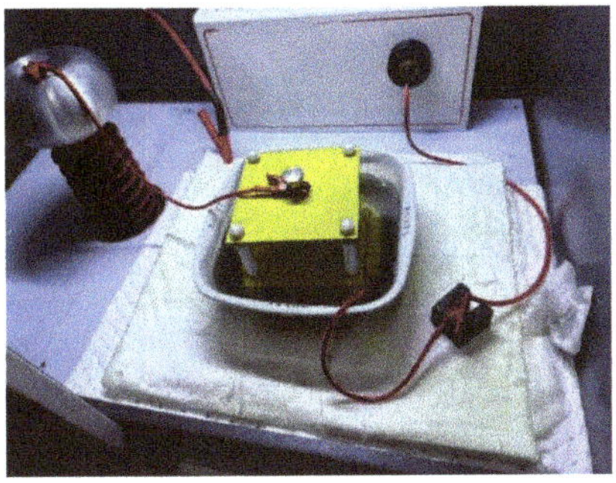

Figure 3. Step-stress method accelerated ageing test pressure test diagram.

3. Measurement Results

The step-up voltage accelerated-ageing test was performed on aged cable insulation sheet samples to record the time required for the voltage level to incrementally increase from the initial voltage according to predefined time-steps until breakdown occurred. The results are shown in Table 1.

When the number of samples in the step-up voltage test is sufficiently large, the total breakdown time can be assumed to follow a Weibull distribution. Figure 4 illustrates the Weibull distribution of breakdown times for XLPE cable slices subjected to step-up voltage ageing tests under different time-steps. In the statistical analysis a confidence level of 0.95 was set. Additionally, the scale parameter of the Weibull distribution is defined as the characteristic ageing time, representing the typical time at which the material is likely to reach breakdown under specific conditions. Using the characteristic ageing times obtained from each experimental group, the corresponding characteristic breakdown voltages were further determined.

Table 1. Gradual voltage rise test data of cable slices in different ageing states.

Sample	Step Duration (s)	No.	Breakdown Voltage (kV)	Total Time (s)	Last Step Time (s)
AS0	200	1	95	2135	90
		2	85	1678	33
		3	90	1920	75
		4	95	2194	149
		5	90	1887	42
		6	85	1762	117
	600	1	80	4519	274
		2	85	4884	39
		3	85	4954	109
		4	80	4337	92
		5	90	5494	49
		6	85	5191	346
	1800	1	80	12,696	51
		2	80	13,725	1080
		3	75	11,186	341
		4	90	12,509	1664
		5	65	10,546	1501
		6	70	9327	282

Table 1. *Cont.*

Sample	Step Duration (s)	No.	Breakdown Voltage (kV)	Total Time (s)	Last Step Time (s)
AS3	200	1	75	1374	129
		2	90	2017	172
		3	95	2187	142
		4	85	1783	138
		5	80	1461	16
		6	85	1679	34
	600	1	90	5459	14
		2	80	4399	154
		3	80	4505	260
		4	80	4336	91
		5	70	3446	401
		6	80	4519	274
	1800	1	70	10,832	1787
		2	65	8100	855
		3	65	8933	1688
		4	80	12,723	78
		5	75	11,645	800
		6	75	11,420	575
AS5	200	1	80	1474	29
		2	80	1550	105
		3	70	1106	61
		4	85	1768	123
		5	80	1528	83
		6	90	1935	90
	600	1	70	3167	122
		2	80	4273	28
		3	65	2522	77
		4	60	1993	148
		5	90	5801	356
		6	85	5165	320
	1800	1	70	9117	72
		2	75	10,873	28
		3	75	11,768	728
		4	65	7381	136
		5	70	9195	150
		6	65	7956	711

Table 2 presents the Weibull distribution parameters of XLPE sheet samples with different ageing degrees under three step durations. The results show that the characteristic breakdown voltage of the same sample decreases with increasing step duration. Under a fixed step duration, XLPE sheet samples with prolonged ageing time exhibited lower characteristic breakdown voltages.

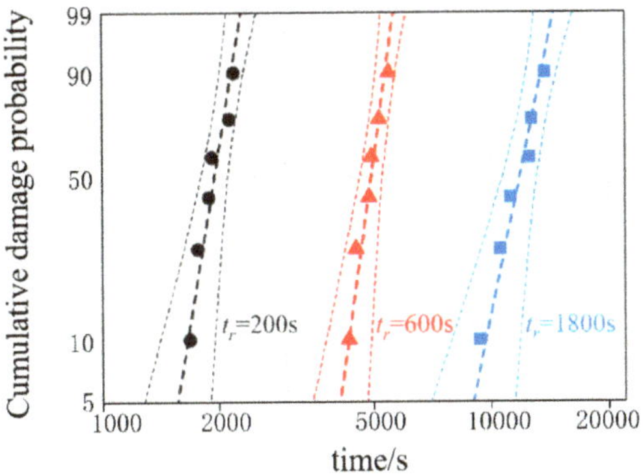

(a) **AS0**

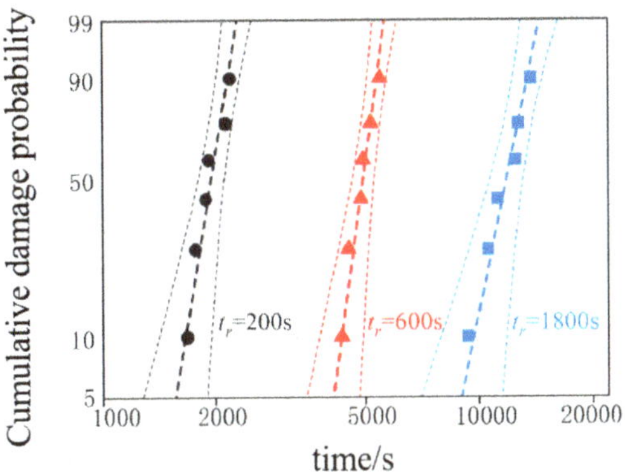

(b) **AS3**

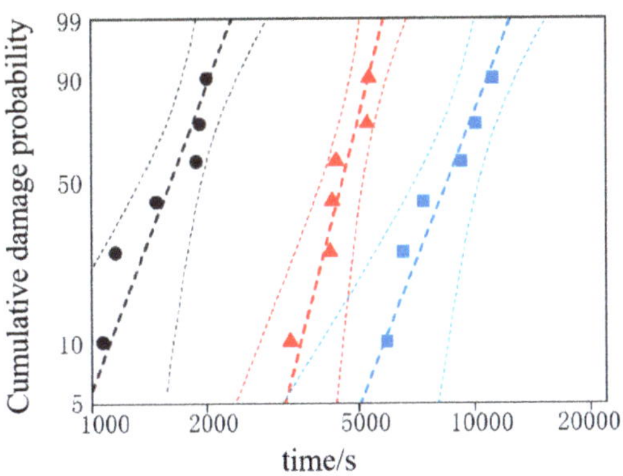

(c) **AS5**

Figure 4. Distribution of breakdown time for XLPE sheet specimens (Black: t_r = 200 s, Red: t_r = 600 s, Blue: t_r = 1800 s).

Table 2. Weibull distribution parameters of cable slices in different ageing states.

Sample	Step Duration (s)	Characteristic Breakdown Voltage (kV)	Characteristic Ageing Time (s)	Last Step Time (s)
Un-aged	200	90	2015	170
	600	85	5077	232
	1800	75	12,308	1463
AS3	200	90	1874	29
	600	80	4701	456
	1800	75	11,284	439
AS5	200	85	1697	52
	600	80	4290	45
	1800	70	10,539	1194

4. Analysis

4.1. Insulation Ageing Lifetime Evaluation Based on Inverse Power Model

The inverse power model, which is the most widely used empirical model in engineering [11,12], can be used to estimate the electrical ageing life of materials. It is expressed by Formula (1) as follows:

$$t_2 = t_1 \left(\frac{U_2}{U_1} \right)^n \tag{1}$$

where n is the ageing exponent of the inverse power model, t_1 is the electrical ageing life at U_1, and t_2 is the electrical ageing life at U_2.

Equation (1) can be expressed as follows:

$$C = U^n t \tag{2}$$

In the inverse power model, C is a constant, and the higher the applied voltage, the shorter the time required for insulation breakdown. Due to the uncertainty in insulation failure time, the ageing life also exhibits uncertainty. Consequently, the insulation breakdown time is treated as a probability function.

Equation (2) can be modified based on the step-up voltage test as follows:

$$C = \sum t_r V_i^n + t_f V_f^n \tag{3}$$

where t_r is step duration/s; V_i is voltage level at each step/kV; t_f is final step time/kV; V_f is breakdown voltage/kV. It should be noted that the value of C is usually regarded as an accumulated ageing 'amount', however, its unit depends on the value of n and it actually has no real physical meaning.

To determine the lifetime exponent n, this study defines three step durations t_{r1}, t_{r2}, and t_{r3}, with the final step voltage levels set as V_{f1}, V_{f2}, and V_{f3}, and the final step times as t_{f1}, t_{f2}, and t_{f3}, Equation (3) can be reformulated as follows:

$$\begin{cases} C_1 = \sum t_{r1} V_i^n + t_{f1} V_{f1}^n \\ C_2 = \sum t_{r2} V_i^n + t_{f2} V_{f2}^n \\ C_3 = \sum t_{r3} V_i^n + t_{f3} V_{f3}^n \end{cases} \tag{4}$$

Since directly solving Equation (4) is challenging, the value of n is predefined to facilitate computation, and the ageing life accumulation parameter C is calculated with this preset value. By taking the logarithm of both sides, we obtain the following:

$$\begin{cases} \ln(C_1) = \ln(t_1) + n\ln(V_1) \\ \ln(C_2) = \ln(t_2) + n\ln(V_2) \\ \ln(C_3) = \ln(t_3) + n\ln(V_3) \end{cases} \tag{5}$$

According to Equation (5), three straight lines are plotted in the $\ln V$-$\ln t$ coordinate system, as shown in Figure 5. These lines correspond to material ageing data under three different step durations. Assuming that they are under the same temperature condition, the data from these varying time intervals maintain consistency with their respective lifetime exponent n and the three lines are parallel to each other. If the ageing life accumulation parameter C differs for each step duration, the intercepts of each line with the coordinate axes will also vary. By defining the true value of the lifetime exponent as n_0, the ageing life accumulation parameter C should remain constant under this value, causing the three lines to coincide. However, due to experimental deviations and data uncertainty, it is generally impossible for the three curves to perfectly overlap. As the value of n gradually approaches the true n_0, the spacing between the curves continuously decreases. Thus, the n value that minimizes this spacing is considered the true lifetime exponent.

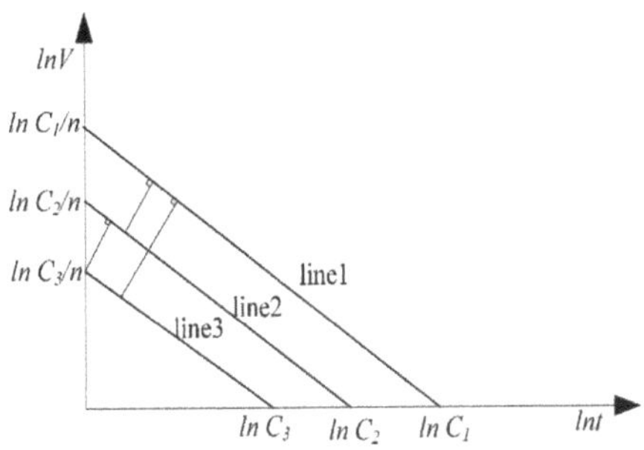

Figure 5. Data analysis in $\ln V$-$\ln t$ coordinate system.

In Figure 5, the sum of the absolute values and the variance of the pairwise distances between the three straight lines can serve as criteria to determine whether the value of n has reached the true value. The distance between any two lines can be expressed by the following formula:

$$d_{12} = \cos\alpha_{12} \cdot \left| \frac{\ln C_1}{n} - \frac{\ln C_2}{n} \right| = \frac{|\ln C_1 - \ln C_2|}{\sqrt{n^2+1}} \tag{6}$$

The following formulas can be used to describe the sum of the distances between straight lines (denoted as *sum*) and the variance of the distances (denoted as *var*):

$$sum = \sum d_{ij} \tag{7}$$

$$\text{var} = \sqrt{\sum d_{ij}{}^2} \tag{8}$$

Within the range of 1–30, we select the n value that yields the lowest *sum* and *var* values as the true value.

The dependence of total inter-curve spacing and variance on the ageing lifetime exponent n in the inverse power model is demonstrated in Figure 6. It can be seen that the values of the *sum* and *var* all decrease first and then increase. There is a lowest point in the figure which is related to the true value of the ageing parameter of the IPM. So, the ageing parameters of samples with different ageing states can be calculated, and the specific data are listed in Table 3.

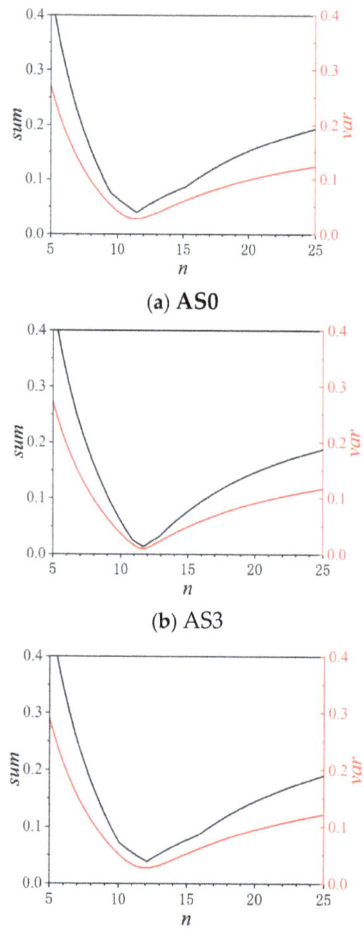

(**a**) **AS0**

(**b**) AS3

(**c**) AS5

Figure 6. The variation in *sum* (black line) and *var* (red line) values of XLPE cable insulation sheet pattern with *n* value.

Table 3. Inverse power model characteristic parameters.

	AS0	AS3	AS5
C_1	1.6×10^{59}	2.67×10^{59}	9.74×10^{60}
C_2	1.13×10^{59}	1.06×10^{60}	9.56×10^{60}
C_3	1.69×10^{59}	4.79×10^{59}	1.04×10^{61}
C	1.47×10^{59}	6.2×10^{59}	9.90×10^{60}
n	11.5	11.7	12.1

It seems that the value of C increases with the ageing time. If it strictly reflects the accumulated ageing 'amount', it suggests that the sample with long ageing time can withstand a higher accumulated ageing 'amount' than the un-aged sample. This does not make sense as the sample with ageing history should have more accumulated ageing amount before the ageing test, and then the tested accumulated ageing 'amount' of the aged sample should be smaller than the un-aged sample. So, here there is a chaos of the physical links between the ageing process and the parameter of IPM.

It can also be seen that the ageing parameter of IPM n increases from 11.5 to 12.1 with the ageing process. Both the values of C and n determine the ageing life curve (the V-t plot), as shown in Figure 7. The ageing parameter n does not have a clear physical meaning, but it defines the slope of the ln$V$-ln$t$, where the bigger the n value the gentler the slope, as shown in Figure 7.

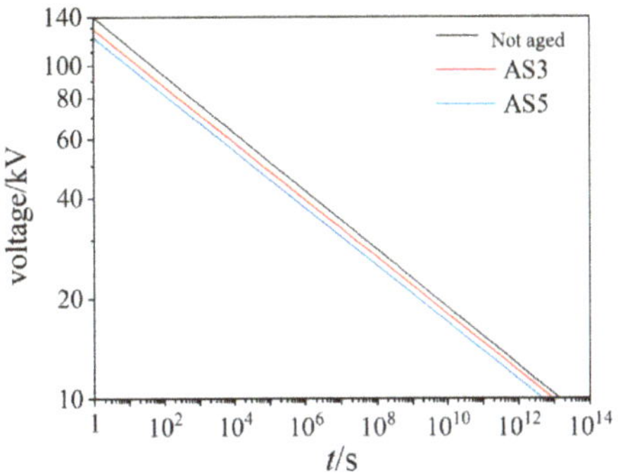

Figure 7. The ageing life curve of samples with different ageing states based on the inverse power model.

It can also be seen that the line for the un-aged sample is above the other lines, with a higher intercept at the Y axis. It shows that the un-aged sample has a higher breakdown voltage than the other samples. The intercept of the X axis at the voltage of 10 kV reflects the estimated lifetime of the sample. It can be seen that the un-aged sample has a longer estimated lifetime than the aged sample.

However, it still does not have a solid physical meaning which can directly link to the physics of the ageing process. And it can hardly explain why the n value increase and how it may link to the insulation property. And it should be noted that the value of C strongly depends on the value of n, as n is the index of V. That is why C cannot truly and strictly reflect the accumulative ageing 'amount'. Both the values of n and C of IPM cannot be used as a parameter to reflect the ageing state of the insulation.

4.2. Insulation Ageing Lifetime Evaluation Based on Crine's Model

The Crine model assumes that polymer ageing is a thermally activated process requiring the overcoming of a free energy barrier, as shown in Figure 8. It considers that the microcavity or free volume is the origin position of the ageing. The ageing happens when the electrons in the free volume obtain higher energy than the ageing activation energy barrier ΔG_0 and break the chain of polymer to make a bigger defect. The applied electric field will increase the energy of the free electrons with an acceleration distance λ. In the assumption of the Crine model, λ is limited by and equal to the maximum length of microcavities or free volume. The variation in activation energy barrier correlates with the material's initial and final free energy states and is influenced by van der Waals bond strength. When the electric field-induced change in activation energy exceeds the interbond cohesive energy, bond rupture occurs. Above the critical field strength, the charges in microcavities have big enough energy to break the chemical chain of the material, leading to a reduction in the insulation property.

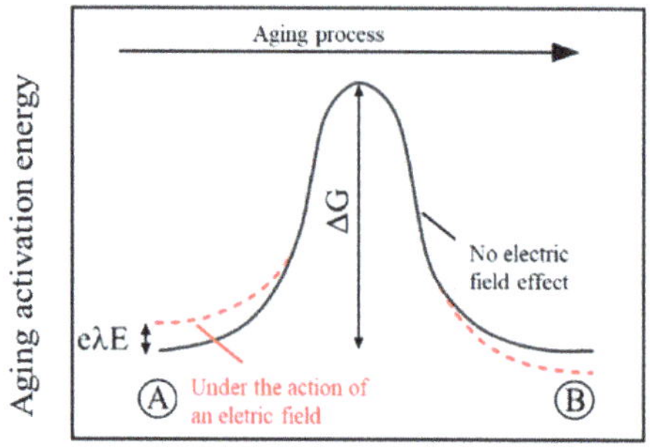

Figure 8. The diagram of the ageing process of the Crine model.

The Crine model formula under high field strength can be expressed as follows:

$$\boldsymbol{t} \approx \frac{h}{2kT} e^{\frac{\Delta G_0 - e\lambda E}{kT}} \quad (9)$$

where T is the absolute temperature/K; k is the Boltzmann constant; h is the Planck constant; E is the electric field strength/V·m^{-1}; ΔG_0 is the activation energy of ageing/eV; and λ is the acceleration distance equal to the maximum length of microcavities or free volume in XLPE/m.

Under constant temperature conditions it is assumed that λ and ΔG_0 remain unchanged. Let $K = \frac{e\lambda}{kTd}$ and $L = \frac{h}{2kT} e^{\frac{\Delta G_0}{kT}}$, which can be considered constants. Equation (9) is rewritten in the following exponential form:

Note: The placeholders "$K = \frac{e\lambda}{kTd}$" and "$L = \frac{h}{2kT} e^{\frac{\Delta G_0}{kT}}$" are retained as in the original text, likely indicating that specific expressions for K and L were omitted.

$$\boldsymbol{t}\boldsymbol{e}^{KU} = L \quad (10)$$

where L can be regarded as a parameter quantifies the accumulative degradation due to ageing. Then, set the test parameters to be consistent with calculating the ageing parameter of the inverse power model. According to the step-stress, Equation (10) can be written in the following form:

$$L = \sum t_i e^{kU_i} + t_f e^{KU_f} \quad (11)$$

where similarly to solving the problem of the ageing parameter for the inverse power model, the values of K are firstly set in a range. And then the values of *sum* and *var* for different K values can be calculation. Then the true value of K can be determined when both *sum* and *var* are with minimum value. Tests conducted over a wide range show that the true value generally falls within the range of 0.0001 to 0.001. Figure 9 illustrates the variation in the *sum* and *var* values with the ageing parameter K for XLPE cable insulation sheet samples at room temperature based on the Crine model.

It can be seen that the values of K decrease first and then increase. It is similar to the condition of Figure 6. The lowest point of K corresponds to the condition in which the three sets of data have the smallest dispersity and is with the true values. Substituting the obtained K and L values into the equation allows for the calculation of acceleration distance λ and ageing activation energy barrier ΔG_0, with the results presented in Table 4.

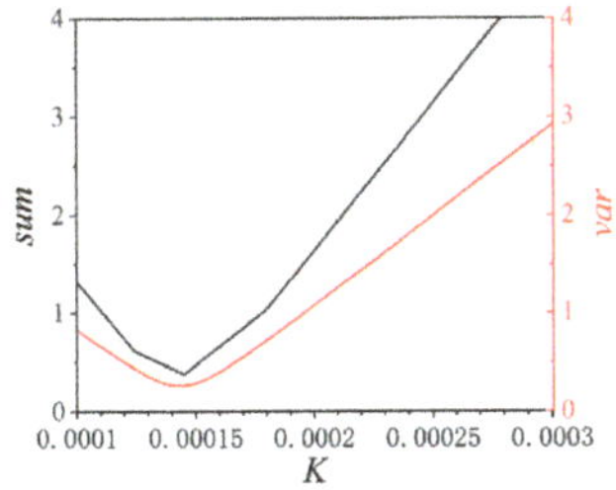

(a) AS0

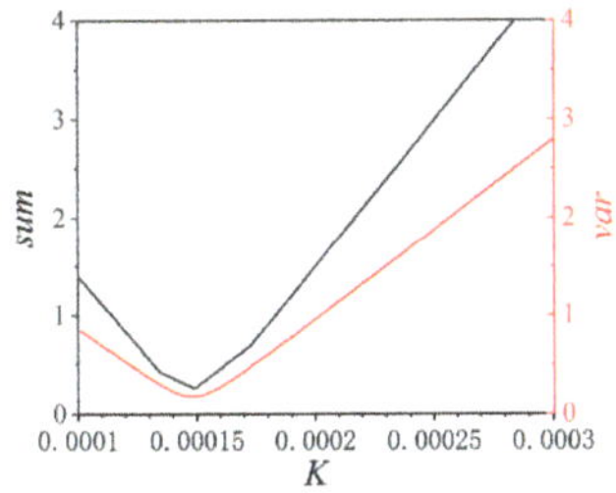

(b) AS3

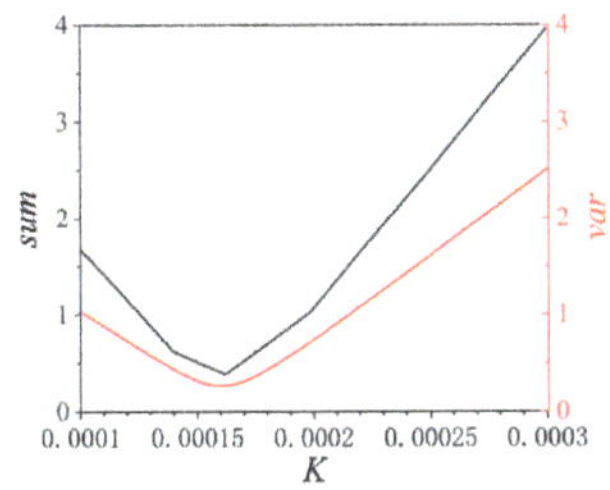

(c) AS5

Figure 9. The variation in *sum* (black line) and *var* (red line) values of XLPE cable insulation sheet pattern with K value.

Table 4. The feature parameters of the Crine model.

	AS0	AS3	AS5
K	0.000145	0.000149	0.00016
L1	1.66×10^8	1.40×10^8	1.49×10^8
L2	1.79×10^8	1.50×10^8	1.93×10^8
L3	1.65×10^8	1.46×10^8	1.93×10^8
L	1.7×10^8	1.45×10^8	1.78×10^8
ΔG_0 (eV)	1.269	1.265	1.27
λ (m)	0.739×10^{-9}	0.753×10^{-9}	0.809×10^{-9}

As shown in Table 4, the ageing activation energy ΔG_0 shows very small variation with the different ageing states. The biggest error between the values of ΔG_0 is just 0.4%. It shows that the ageing activation energy ΔG_0 does not change with the ageing process and can be considered as a material intrinsic parameter. The value of the ageing activation energy ΔG_0 may corresponds to the bond energy of PE chain, the cross-linking network, and the band energy of the XLPE insulation. However, more work is still needed to reveal this detailed relationship.

λ is the acceleration distance of the free electron, and equals the maximum length of microcavities or free volume. Table 4 demonstrates that as ageing time increases, the value of λ increases, leading to greater electron acceleration distances within the microcavities or free volume. It consequently makes the electrons with higher energy more likely to cause damage on XLPE chains. Then the lifetime of the aged sample becomes shorter than the un-aged sample. The test results also make sense as the ageing process may break the polymer chain of XLPE, and make larger sized microcavities or free volume. So the value of the acceleration distance can be regarded as an indicator for the estimation of the insulation lifetime. In the case of unchanged ageing activation energy, the bigger the acceleration distance, the shorter the insulation lifetime.

The ageing lifetime curve based on the Crine model is shown in Figure 10. It can be seen that the ageing lifetime gradually decreases with increasing ageing time, and the remaining lifetime of aged specimens is significantly shorter than that of un-aged specimens. The result is similar to the trend of the IPM, but still has some differences.

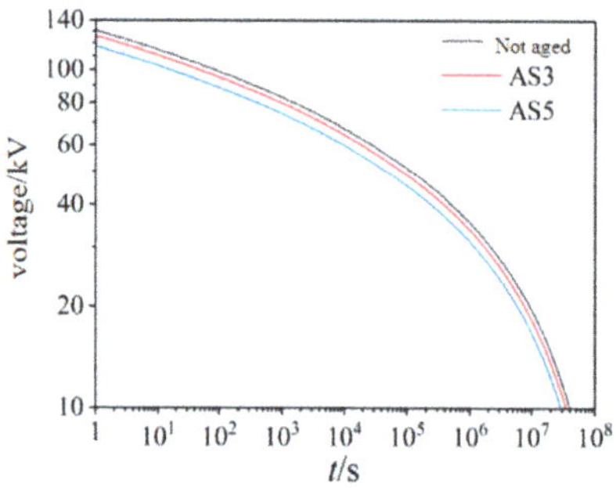

Figure 10. The ageing life curve of samples with different ageing states based on the Crine model.

The first difference is that the slope of $\ln V$-$\ln t$ changes in the ageing life curve based on the Crine model rather than the IPM. The reason is that the higher the electric field, the higher energy the electrons are supposed to observe from the electric field. And then the electrons are more probable to cause damage to the sample in the higher electric field than in the lower electric field. However, due to the existence of the activation energy, it behaves differently with the IPM. When the lifetime is very short, the voltage should be very near to the breakdown voltage. It is more reliable for the Crine model, as the voltage changes slightly with the lifetime when the lifetime is shorter than 10 s. But the IPM suggests that the voltage may continuously increase as there is a decrease in the lifetime. It does not make sense. Similarly, when the voltage is very low, the lifetime of insulation based on the Crine model tends to reach a maximum value rather than continuously increase, as suggested by the IPM. This is also due to the limited activation energy. The electrons cannot only obtain energy from electric field but also from the thermal kinetic activities. They still have a very small possibility to cause damage in the condition of limited activation energy. So, the Crine model describes a more reasonable lifetime trend than the IPM.

The second difference is that the Crine model predict a much shorter lifetime than the IPM at the voltage of 10 kV. The 5 weeks aged samples have about 1.29 years remaining lifetime in the Crine model, while they have a 142,694 years lifetime in the IPM. Even though it is in the condition of room temperature, it is still inaccurate that the 5 weeks aged sample has such a long remaining lifetime. However, it is still hard to clearly figure out which prediction is more reliable, as the long-term test has many factors which are hard to control and takes too much time and cost.

With the discussions above, it can be seen that the key parameters of the Crine model have clear physical meaning, while the IPM can only fit the mathematic pattern of the data. The activation energy barrier of the Crine model does not change with the ageing process, so that it can be considered as a material intrinsic parameter that is determined by the nature of the bond, cross-linking network, and band energy of the XLPE. The acceleration distance of the Crine model increases with the ageing time, so it can be regarded as an indicator of the ageing state. Conversely, the key parameters of the IPM do not have clear physical meaning and their change with ageing time cannot be sufficiently used as an indicator of ageing states. Furthermore, the Crine model describes a more reasonable ageing lifetime pattern than the IPM, especially when the voltage is high and low. So, the Crine model has a greater potential to be used in the remaining lifetime estimation of the cable insulation or other materials.

5. Conclusions

This paper tests the remaining lifetime of aged and un-aged XLPE samples with the step-stress method. The inverse power model and the physical-driven Crine model are employed in the analysis, and the calculation method of the key parameters of the two models are proposed. The characteristics of remaining lifetime and key ageing parameters of the samples with different ageing states are discussed regarding the two models. The following conclusions are drawn:

1. The step-stress method can efficiently obtain the lifetime data of the insulation samples. It is found that the longer the ageing time, the shorter the remaining lifetime of the XLPE sample.
2. The ageing parameters n and C of the IPM increase with the ageing time, but they can hardly link to the physics of the ageing process, and they cannot be used as an indicator of the ageing state.
3. The activation energy barrier of the Crine model does not change with the ageing process, and so it can be considered as a material intrinsic parameter determined by the material nature. The acceleration distance of the Crine model increases with the ageing time, and it can be regarded as an indicator of the ageing state.
4. The Crine model describes a more reasonable ageing lifetime pattern than IPM, especially when the voltage is high and low. The Crine model has a greater potential to be used in the remaining lifetime estimation of the cable insulation or other materials.

Author Contributions: Methodology, J.M.; Investigation, Y.S., J.W. and J.X.; Data curation, Y.S. and J.Q.; Writing—original draft, Y.S. and J.C.; Writing—review & editing, Z.L.; Visualization, Y.L. All authors have read and agreed to the published version of the manuscript.

Funding: This paper is funded by the Science and Technology Project of State Grid Beijing Electric Power Company. The project name is "Research and Application of Cable Insulation Defect Detection and Ageing Assessment Technology Based on Frequency Domain Dielectric Spectroscopy," and the project code is 520246230008.

Data Availability Statement: The original contributions presented in this study are included in the article. Further inquiries can be directed to the corresponding author.

Conflicts of Interest: Authors Yingqiang Shang, Jingjiang Qu, Jingshuang Wang and Jun Xiong were employed by the company State Grid Beijing Electric Power Co., Ltd. The remaining authors declare that the research was conducted in the absence of any commercial or financial relationships that could be construed as a potential conflict of interest.

References

1. Du, B.X.; Li, Z.L.; Yang, Z.R.; Li, J. Application and Research Progress of HVDC Cross-Linked Polyethylene Cables. *High Volt Eng.* **2017**, *43*, 344–354.
2. Zhou, T.X.; Wang, L.Y.; Ji, X.P.; Han, X.Y. Power Grid Scale Prediction and Optimization Needs Based on Policy Requirements of Modern Infrastructure System. *Electr. Power* **2024**, *57*, 1–8.
3. China Electricity Council. *Annual Development Report of China's Electric Power Industry 2019*; China Building Materials Industry Press: Beijing, China, 2019.
4. Dakin, T.W.; Studniarz, S.A. The voltage endurance of cast and molded resins. In Proceedings of the 1977 EIC 13th Electrical/Electronics Insulation Conference, Chicago, LI, USA, 25–29 September 1977.
5. Endicott, H.S.; Hatch, B.D.; Sohmer, R.G. Application of the eyring model to capacitor ageing data. *IEEE Trans. Compon. Parts* **1965**, *12*, 34–41. [CrossRef]
6. Lewis, T.J.; Liewellyn, J.P.; Freestone, J.; Hampton, R.N. A newmodel for electrical ageing and breakdown in dielectrics. In Proceedings of the Seventh International Conference on Dielectric Materials, Measurements and Applications, IET, Bath, UK, 23–26 September 1996.
7. Dissado, L.; Mazzanti, G.; Montanari, G.C. The incorporation of space charge degradation in the life model for electrical insulating materials. *IEEE Trans. Dielectr. Electr. Insul.* **1995**, *2*, 1147–1158. [CrossRef]
8. Crine, J.P.; Parpal, J.L.; Lessard, G. A model of ageing of di-electric extruded cables. In Proceeding of the 3rd International Conference on Conduction and Breakdown in Solid Dielectrics, Trondheim, Norway, 3–6 July 1989.
9. *GB/T 11026.1-2012*; National Standard of the People's Republic of China. Electrical Insulating Materials—Thermal Endurance—Part 1: Ageing Procedures and Evaluation of Test Results. General Administration of Quality Supervision, Inspection and Quarantine of the People's Republic of China and the Standardization Administration of the People's Republic of China: Beijing, China, 2012.
10. Zhang, Z.; Li, Y.; Zheng, Y.; Li, H.; Zhao, X.; Li, B.; Pei, X.; Cheng, Y. Insulation Aging Assessment Method for HVDC Cable Based on Space Charge Characteristics. *Guangdong Electr. Power* **2019**, *32*, 98–105.
11. Wang, Y.; Lv, Z.; Wang, X.; Wu, K.; Zhang, C.; Li, W.; Dissado, L.A. Estimating the inverse power law aging exponent for the DC aging of XLPE and its nanocomposites at different temperatures. *IEEE Trans. Dielectr. Electr. Insul.* **2016**, *23*, 3504–3513.
12. Xia, W.; Xiaotong, S.; Quanyu, L.; Wu, K.; Tu, D. Research progress on insulation ageing life model based on space charge effect. *High Volt. Eng.* **2016**, *42*, 861–867.

energies

MDPI

Article

The Molecular Dynamics of Signature Gas Diffusions in Synthetic-Ester-Based Oil Under a Range of Thermal Conditions

Liping Guo [1], Hongliang Wang [1], Weiwei Qi [2], Jun Zhang [2] and Wu Lu [2,*]

1 Shanghai Electric Transmission and Distribution Experimental Center Co., Ltd., Shanghai 200072, China
2 College of Electrical Engineering, Shanghai University of Electric Power, Shanghai 200090, China
* Correspondence: wuluee@shiep.edu.cn

Abstract

Synthetic ester insulating oils are extensively utilized in power transformers due to their exceptional insulating properties, thermal stability, and environmental compatibility. The dissolved gas analysis (DGA) technique, which is employed to diagnose internal faults in transformers by monitoring the concentration and composition of dissolved gases in oil, is thought to be effective in detecting typical faults such as overheating and partial discharges in synthetic esters. However, owing to the significant differences in the properties of traditional mineral oil and synthetic esters, the existing DGA-based diagnostic methods developed for mineral oils cannot be directly applied to synthetic esters. A deep understanding of the microscopic processes occurring during the gas generation and diffusion of synthetic esters is an urgent necessity for DGA applications. Therefore, in this study, we systematically investigated the diffusion behavior of seven typical fault gases in synthetic ester insulating oils within a temperature range of 343–473 K using molecular dynamics simulations. The results demonstrate that H_2 exhibits the highest diffusion capability across all temperatures, with a diffusion coefficient of 33.430×10^{-6} cm^2/s at 343 K, increasing to 402.763×10^{-6} cm^2/s at 473 K. Additionally, this paper explores the microscopic mechanisms underlying the diffusion characteristics of these characteristic gases by integrating the Free-Volume Theory, thereby providing a theoretical foundation for refining the fault gas analysis methodology for transformer insulating oils.

Keywords: synthetic ester; dissolved gas analysis; molecular dynamics; diffusion coefficient; free volume

1. Introduction

As the global energy system is showing a low-carbon and environmentally sustainable trend, the demand for sustainable and high-performance insulating materials for power equipment is becoming increasingly critical. Synthetic ester insulating oils, characterized by their high ignition points, biodegradability, excellent dielectric strength, and thermal stability, have emerged as promising alternatives to traditional mineral oils, particularly for power transformers with stringent environmental requirements as well as high-voltage and large-capacity applications, offering significant advantages [1,2]. Despite the excellent performance of synthetic esters in terms of environmental friendliness and thermal stability, the characteristic gas diffusion mechanism of synthetic ester oils under the influence of typical internal discharge or overheating faults in transformers is still unclear, which has led to the challenge of fault diagnosis based on the dissolved gas analysis (DGA) technique in oil. Furthermore, the diagnostic criteria of traditional DGA methods are based on mineral oil systems, whereas the diffusion behavior and equilibrium characteristics of dissolved gases in synthetic esters may differ markedly from those in mineral oils due to inherent differences in viscosity, polarity, and molecular structure. Direct application of the existing diagnostic criteria may lead to underestimation or misjudgment of gas concentration thresholds and increase the risk of fault misdiagnosis due to discrepancies in diffusion kinetics. Therefore, there is an urgent need to elucidate the diffusion mechanisms of characteristic gases in synthetic esters, as doing so would contribute to the refinement and optimization of fault diagnosis methods for synthetic-ester-immersed transformers based on DGA technology [3–6].

Recent research on synthetic esters has predominantly focused on optimizing their electrical properties, thermal aging characteristics, and compatibility with solid insulating materials. Studies have demonstrated that synthetic esters exhibit superior dielectric strength and lower dielectric loss compared to mineral oils, along with a slower increase in acid value during prolonged thermal aging [7,8]. Additionally, synthetic esters show enhanced compatibility with cellulose insulating paper, which helps mitigate the aging of an insulating material [9,10], and their oxidative stability at elevated temperatures is significantly superior to that of mineral oils [11]. However, most of the existing studies have concentrated on macroscopic physical parameters (e.g., dielectric constant, viscosity, etc.) and analyses of aging products, with limited attention given to the systematic exploration of the diffusion kinetics of dissolved gases within the synthetic ester system [12,13]. Although a few experimental investigations have compared the gas solubility differences between synthetic esters and mineral oils, the experimental conditions often hinder the elucidation of the diffusion mechanism at the molecular level, resulting in an incomplete understanding of migration patterns [14].

In the DGA technique, internal transformer faults are diagnosed by monitoring the concentration and proportion of dissolved gases in the oil, and the technique's effectiveness relies on an accurate knowledge of the gas diffusion mechanism [15]. At present, the traditional DGA technique based on mineral oil is highly mature, and its diagnostic logic and gas concentration thresholds have been fully verified through long-term engineering practice, but such methods are not directly applicable to the fault analysis of natural ester insulating oils, and it is doubtful whether this technique's theoretical framework is applicable to synthetic ester systems [16,17]. In view of the characteristics of natural ester insulating oils, some scholars preliminarily explored the DGA diagnostic framework adapted to natural esters by experimentally modifying the gas solubility parameter and adjusting the threshold value of the proportion of fault gases [18–20]. However, despite some scholars' attempts to experimentally study the macroscopic characteristics of synthetic esters, there is a lack of studies determining their characteristic gas diffusion behaviors and gas diffusion kinetic mechanisms. These bottlenecks have constrained the development of DGA diagnostic criteria applicable to synthetic esters, and there is an urgent need to reveal the gas diffusion mechanisms of synthetic esters at the molecular dynamics level [21–23].

To address the above issues, we employed molecular dynamics simulations to systematically investigate the diffusion behaviors of seven typical fault-characteristic gases in synthetic ester insulating oil within the temperature range of 343 K to 473 K. Utilizing the molecular dynamics simulation approach, a synthetic ester molecular model and a gas miscible system were developed. This study further examines the influence mechanism of temperature gradients on the gas diffusion behavior of the insulating oil at the microscopic level through the calculation of mean square displacements (MSDs) and gas diffusion coefficients in conjunction with the free-volume theory. The findings of this research can be applied to the design optimization of DGA technology for synthetic ester insulating oils, thereby facilitating the reliable operation and intelligent maintenance of environmentally friendly transformers.

2. Simulation Methodology

2.1. Molecular Diffusion Modeling of Synthetic Esters

We constructed a molecular dynamics model of a synthetic ester composed of Pentaerythritol tetranonanoate (PENTA-TNA, $C_{41}H_{76}O_8$) and its composite system with fault-characteristic gases. The structure of a single synthetic ester molecule consists of a pentaerythritol backbone and four isononyl ester groups, which confer excellent thermal stability and outstanding electrical properties. To accurately characterize the macroscopic properties of this system, an initial model was constructed using 10 PENTA-TNA molecules, and periodic boundary conditions were applied to eliminate size effects. Furthermore, multi-stage dynamic relaxation and geometric optimization were performed to eliminate potential atomic-level anomalous contact and high-energy configurations in the initial model, facilitating the model's stabilization to a reasonable state. Specifically, the steps include achieving a 100 ps thermal equilibrium under the NPT ensemble, realizing 100 ps energy stabilization under the NVE ensemble, pressure adjustment under the low-pressure NPT ensemble, and 10,000-step geometric optimization based on the conjugate gradient method. After carrying out these optimization steps, the final PENTA-TNA molecular model, with a unit cell length of 23.28 Å and a density of 0.917 g/cm^3, was obtained, as shown in Figure 1.

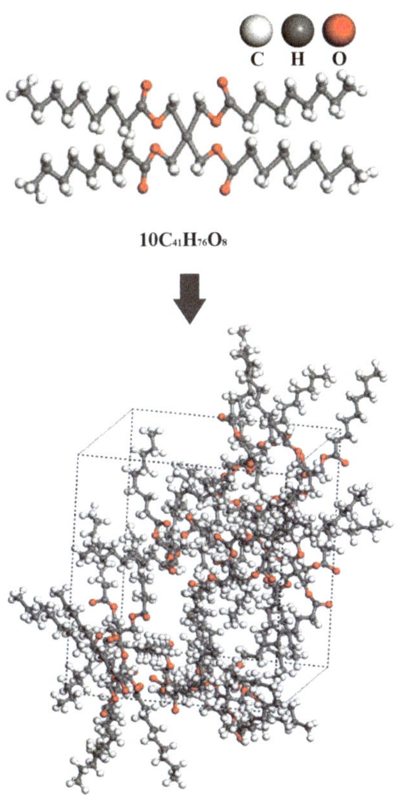

Figure 1. Molecular modeling of a synthetic ester.

In this study on the diffusion behavior of dissolved gases in oil, a composite molecular model containing seven characteristic gases was constructed based on the PENTA-TNA base oil model. During the construction of the molecular models for small gas molecules using Materials Studio, small gas molecules and synthetic ester molecules were differentiated by color: C in the small gas molecules was colored dark gray, H was given in emerald green, and O was given in light blue. As shown in Figure 2, these molecular models are complete and strictly adhere to the standard requirements for initial configurations in molecular dynamics simulations. In this study, 10 gas molecules were randomly embedded in the base oil model formed by 10 PENTA-TNA molecules, and the overall density was set to 0.95 g/cm^3 to match the physical property requirements under actual operating conditions. As exemplified by the CH_4 diffusion model shown in Figure 3, this system maintains the stable configuration of the base oil while ensuring accurate description of short-range interactions, such as van der Waals forces, through reasonable molecular spacing.

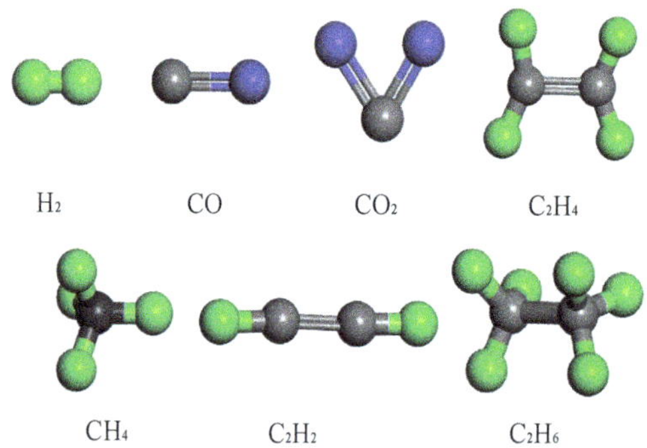

Figure 2. Gas molecule modeling.

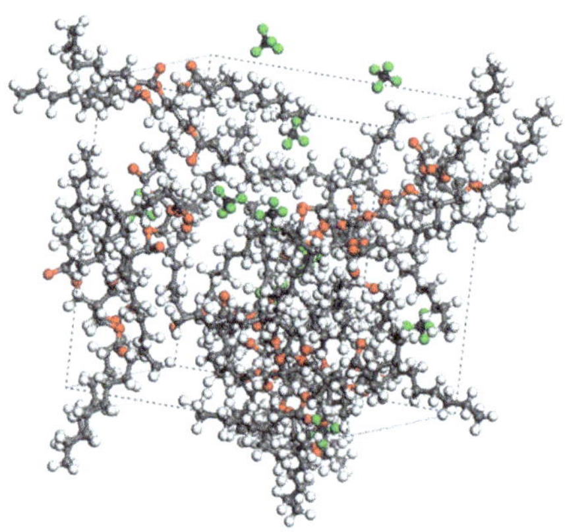

Figure 3. Molecular modeling of mixed oil-and-gas system. (Hydrogen and carbon atoms in CH_4 are shown in green and gray, respectively, and oxygen, hydrogen, and carbon atoms in the molecule are shown in red, gray, and white, respectively).

2.2. Simulation Parameter Settings

We constructed a computational framework based on the PCFF force field to simulate the diffusion behavior of dissolved gases in synthetic esters using molecular dynamics. The force field builds upon the core parameters of the CFF91 force field and significantly enhances simulation accuracy for complex organic molecular systems by extending the functional expressions for bond length vibration, bond angle bending, dihedral torsion, and non-bond interactions in polymeric systems. To address the non-bond interactions between the synthetic ester and gas molecules, the van der Waals forces were computed using the Atom-Based truncation method, which reduces computational load within a reasonable range and improves simulation efficiency. The electrostatic interactions are handled using the Ewald summation method, which accurately describes long-range Coulomb forces while also maintaining computational efficiency.

Additionally, in terms of thermodynamic control algorithms, a hierarchical multi-level control approach was applied. Firstly, temperature regulation was achieved through the Andersen thermostat algorithm, whose random collision mechanism effectively maintained the thermal equilibrium of the system and prevented local thermal anomalies. Secondly, pressure control was carried out using the Berendsen method, ensuring stable convergence of the system under a standard pressure of 0.1 MPa, thereby simulating the actual working environment of transformers. Lastly, the Velocity Verlet algorithm was employed for dynamic integration, balancing energy conservation with trajectory accuracy to ensure the authenticity of molecular motion behavior. The three-dimensional system constructed during the simulations mitigates finite-size effects by employing periodic boundary conditions, which enables a more accurate simulation of molecular diffusion behavior in real oils [24,25].

Before conducting molecular dynamics simulations, it is essential to optimize the geometric structure of a system and relax the pressure. Specifically, the process is divided into the following steps:

(1) A preliminary composite model of synthetic ester oil molecules and gas molecules is constructed and undergoes 10,000 steps of geometric structure optimization to eliminate potential atomic overlaps and non-physical contact.

(2) Pressure relaxation is performed under the NPT ensemble for 100 ps to allow the system's density to converge to 0.95 g/cm^3, ensuring the accuracy of subsequent simulations.

After the above steps were completed, a 600 ps molecular dynamics simulation was carried out under the NVT ensemble conditions. The simulation process was divided into two parts: the first 100 ps were used to achieve a stable molecular state, and the remaining 500 ps were used for data collection to analyze the diffusion characteristics of dissolved gas molecules. The recording interval of the molecular trajectory was reduced to 1 ps in order to capture the complete gas diffusion process, providing a reliable data foundation for the calculation of diffusion coefficients.

Furthermore, considering that the actual operating temperature of transformer insulating oil typically ranges from 313 K to 423 K, four simulation temperatures were selected to represent different operating conditions pertaining to the transformer: 343 K represents the normal operating state; 393 K represents the temperature under high load or mild overheating conditions; 423 K corresponds to the early stage of thermal failure; and 473 K represents a high temperature close to the point of discharge failure. By performing molecular dynamics simulations under these temperature conditions, the diffusion coefficients of each characteristic gas in synthetic ester were calculated, and the micro-mechanisms of the effect of temperature on the diffusion characteristics of the gases were analyzed in combination with the free volume, which provides data support for the optimization of DGA technology.

3. Results and Discussion

3.1. Characteristic Gas Diffusion Properties

In the synthetic ester system, the diffusion characteristics of gas molecules of different types exhibit significant differences. This study utilizes the mean squared displacement (MSD) curve to clearly describe the diffusion characteristics of gases in the oil and their correlations with temperature. According to molecular dynamics theory, the mean squared displacement is defined as the statistical average of the deviation of the positions of all particles in the system relative to a reference position over a time interval t. The expression is given below:

$$MSD(t) = \left\langle \left| r(t) - r(0) \right|^2 \right\rangle \quad (1)$$

where *r*(*t*) is the particle's position at a given time, *r*(0) is the initial reference position, and $\langle X \rangle$ represents the statistical average within the system.

Figure 4 presents a schematic of the mean squared displacement (MSD) for various characteristic gases, depicted using a log–log dependence. Of all the gasses, H_2 demonstrates the strongest diffusion ability under all temperature conditions, maintaining the highest mean squared displacement (MSD) compared to the other gases. This is primarily attributed to the minimal molecular weight of H_2, provoking the least resistance from oil molecules, as well as the pronounced effect of temperature increases on enhancing its diffusion capacity. As depicted in Figure 4a,b, at 343 K, the MSD of H_2 is an order of magnitude higher than that of the other gases, and as the temperature rises to 473 K, its diffusion advantage remains evident. This reflects the highly efficient diffusion characteristics of lightweight non-polar molecules in synthetic esters. Additionally, in the later stages of hydrogen diffusion, deviations from the mean squared displacement (MSD) curve can be observed under varying temperature conditions. For instance, as shown in Figure 4a, the tail of the curve exhibits a sharp upward trend, whereas in Figure 4d, the tail demonstrates a sudden decline. This phenomenon can be attributed to two factors: first, the exceptionally high diffusion ability of H_2 molecules, which makes them more susceptible to thermal fluctuations or boundary effects in small-scale systems, and, second, the limited number of particles within the simulated system, making the system prone to inducing statistical fluctuations, particularly for highly diffusive gases.

For the gases CO and CO_2 oxide, the MSD of CO_2 is consistently higher than that of CO, with the difference being particularly noticeable at lower temperatures. However, as the temperature increases, the diffusion capabilities of the two gases gradually converge. For example, at 473 K, the MSD of CO_2 is only slightly higher than that of CO. This phenomenon could be attributed to the increased thermal motion of oil molecules at higher temperatures, which reduces the system's viscosity and thus diminishes the impact of molecular weight differences on diffusion. Furthermore, the diffusion processes of CO and CO_2 are primarily dominated by the physical properties of the insulating oil, as their polar nature leads to lower solubility in weakly polar synthetic ester oils.

In addition, the diffusion characteristics of hydrocarbon gases exhibit complex temperature sensitivity. At 343 K, C_2H_2 has the highest MSD, indicating that, under this temperature condition, its diffusion ability is the strongest among hydrocarbon gases. This may be related to the lower spatial hindrance resulting from its linear molecular structure, as shown in Figure 4a. As shown in Figure 4b, when the temperature rises to 393 K, C_2H_4 exhibits the highest MSD among hydrocarbon gases, likely due to its double-bond structure, which allows intermolecular forces to be overcome at higher temperatures. Figure 4c,d reveal that CH_4 shows a significant enhancement in diffusion ability at higher temperatures (423–473 K), possibly due to the decrease in synthetic ester viscosity during the temperature rise, with the simple spherical molecular structure of CH_4 benefiting the most from this decrease. However, C_2H_6 consistently demonstrated the slowest diffusion rate among the hydrocarbon gases, likely due to the inhibitory effect of its greater molecular weight and branched structure. This variation in diffusion order under different temperature conditions suggests that the diffusion characteristics of hydrocarbon gases are not only closely related to temperature but also influenced by the molecular structure of each gas, intermolecular interactions, and solubility in the oil.

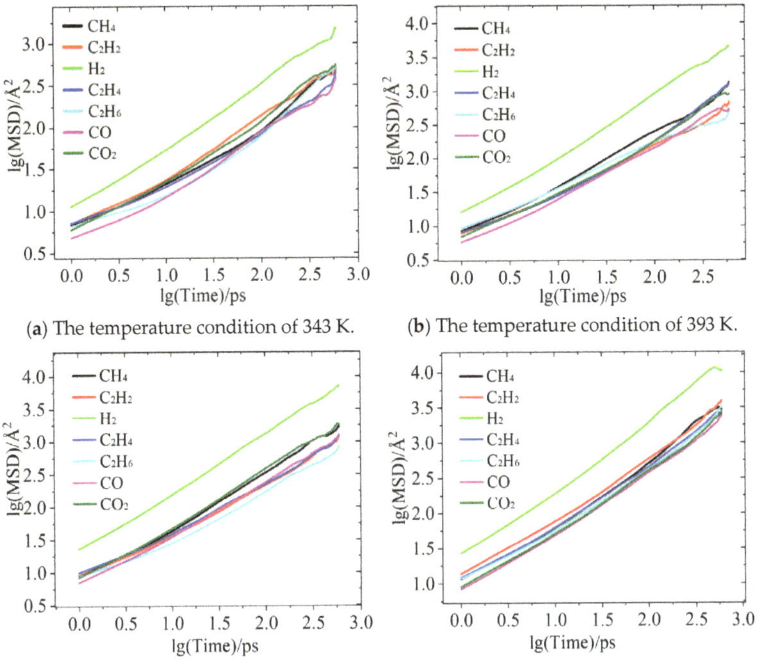

(a) The temperature condition of 343 K. (b) The temperature condition of 393 K.

(c) The temperature condition of 423 K. (d) The temperature condition of 473 K.

Figure 4. MSD plots of diffusion behavior of characteristic gases.

Overall, the diffusion abilities of characteristic gases align with the following trend: H_2 > hydrocarbon gases > CO and CO_2. However, the relative diffusion capabilities fluctuate dynamically with temperature. Moreover, the diffusion characteristics are determined by factors such as molecular weight, polarity, molecular structure, and the properties of insulating oils. The exceptional diffusion ability of H_2 is primarily attributed to its extremely low molecular weight, while the complex temperature-dependent behavior of hydrocarbon gases reveals the influence of molecular structure on the diffusion process. The findings of this study provide an important theoretical basis for optimizing gas-monitoring technologies based on diffusion characteristics.

3.2. Characteristic Gas Diffusion Coefficient

The diffusion coefficient is a key parameter for quantitatively describing the migration ability of gas molecules in insulating oils, and it is calculated based on the linear relationship between the mean squared displacement (MSD) and time. By analyzing the MSD curves in Figure 4, it can be observed that within the time range of 25 ps to 500 ps, the MSD of all the gases exhibits a strong linear relationship with time, indicating that molecular motion is primarily governed by free diffusion during this period. To accurately calculate the diffusion coefficient (D), it is necessary to select the linear segment of the MSD curve for linear regression fitting. The slope (k) obtained from this fitting is then used to establish a quantitative relationship with the diffusion coefficient:

$$D = \frac{k}{6} \quad (2)$$

where the coefficient 6 indicates that the diffusion system is three-dimensional.

Additionally, the goodness of fit for the linear regression is represented by the correlation coefficient R^2, and it can be calculated using the following formula:

$$R^2 = 1 - \frac{\sum (l_i - \hat{l}_i)^2}{\sum (l_i - \bar{l})^2} \tag{3}$$

where l_i represents the actual MSD values; $\hat{l}_i$ represents the fitted MSD values; and $\bar{l}$ represents the mean MSD values.

Based on the aforementioned method and by combining the MSD values of the characteristic gases, the diffusion coefficients of seven characteristic gases in the synthetic ester were calculated at four different temperatures: 343 K, 393 K, 423 K, and 473 K. The results reveal the temperature-dependent behavior of gas diffusion and its correlation with molecular characteristics. As shown in Table 1, the diffusion coefficients of the characteristic gases significantly increased when increasing the temperature, with H_2 exhibiting the most notable diffusion capacity. At 343 K, the diffusion coefficient of H_2 was 33.430×10^{-6} cm^2/s, and as the temperature increased, it rose to 402.763×10^{-6} cm^2/s at 473 K, marking a 12-fold increase, far exceeding that of the other gases. This can be attributed to the extremely low molecular weight of H_2 and its weak intermolecular forces, making it highly sensitive to temperature changes.

Table 1. Characteristic gas diffusion coefficients at four temperatures.

		H_2	CO	C_2H_2	CH_4	CO_2	C_2H_6	C_2H_4
343 K	Slope, k	2.006	0.434	0.799	0.862	0.943	0.918	0.539
	Fitted correlation coefficient, R^2	0.976	0.980	0.987	0.991	0.973	0.998	0.990
	Diffusion coefficient $D/\times10^{-6}$ cm^2/s	33.430	7.225	8.992	14.365	15.722	15.295	8.992
393 K	Slope, k	6.936	1.087	2.010	1.627	1.786	0.511	2.010
	Fitted correlation coefficient, R^2	0.990	0.994	0.994	0.990	0.997	0.926	0.999
	Diffusion coefficient $D/\times10^{-6}$ cm^2/s	115.607	18.113	13.915	27.112	29.770	8.510	33.494
423 K	Slope, k	10.911	1.804	1.809	2.665	2.656	1.114	1.685
	Fitted correlation coefficient, R^2	0.994	0.994	0.993	0.972	0.981	0.988	0.992
	Diffusion coefficient $D/\times10^{-6}$ cm^2/s	181.85	30.074	30.154	44.410	44.263	18.559	28.088
473 K	Slope, k	24.166	3.377	5.983	6.187	3.998	4.411	4.741
	Fitted correlation coefficient, R^2	0.999	0.998	0.992	0.995	0.990	0.997	0.997
	Diffusion coefficient $D/\times10^{-6}$ cm^2/s	402.763	56.278	99.719	103.119	66.631	73.524	79.014

Hydrocarbon gases such as CH_4, C_2H_2, C_2H_4, and C_2H_6 also exhibited significant increases in diffusion coefficients at 473 K compared to the values at 343 K, surpassing 70×10^{-6} cm^2/s. For example, the diffusion coefficient of CH_4 increased from 14.365×10^{-6} cm^2/s to 103.119×10^{-6} cm^2/s. Among these gases, C_2H_2 showed the most significant temperature-dependent diffusion change, with its diffusion coefficient increasing from 18.992×10^{-6} cm^2/s to 99.719×10^{-6} cm^2/s, indicating that its linear molecular structure allows for more efficient diffusion into the insulating oil at higher temperatures. In contrast, oxide gases such as CO and CO_2 exhibited lower diffusion coefficients than H_2 and hydrocarbons but still showed a steady increasing trend. The diffusion coefficient of CO_2 increased from 15.722×10^{-6} cm^2/s at 343 K to 66.631×10^{-6} cm^2/s at 473 K, while that of CO increased from 7.225×10^{-6} cm^2/s to 56.278×10^{-6} cm^2/s, with growth factors of 4.2 and 7.8, respectively.

Furthermore, the behavior of the diffusion process can be further characterized by the slope, k. The data shows that the diffusion behavior of the characteristic gases intensifies with an increase in temperature. It is noteworthy that the slope for C_2H_6 exhibited an unusual decrease at 393 K (0.511) but recovered and resumed growth at higher temperatures, suggesting that the molecular interactions or solubility of C_2H_6 might change in this specific temperature range, temporarily inhibiting its diffusion. The reliability of the diffusion model was verified through the fitting correlation coefficient (R^2), particularly at 473 K, where all the gases exhibited R^2 values above 0.99, confirming that the constructed diffusion model effectively describes the diffusion characteristics of the characteristic gases.

In conclusion, the relative strength of gas diffusion and its temperature response characteristics provide a reference for the dynamic monitoring of dissolved gases in transformer insulating oil. For example, the highly sensitive diffusion characteristics of H_2 at high temperatures highlight its potential advantage in early fault warning. These findings foster a deeper understanding of the diffusion mechanisms of gases in insulating oil from a molecular dynamics' theory, laying a theoretical foundation for optimizing DGA-based detection technologies based on diffusion characteristics.

3.3. Analysis of the Free Volume

The Free-Volume Theory (FVT) is an important theoretical model used to explain the diffusion behavior of molecules in liquids or polymers. Its core assumption is that there are unoccupied spaces within a liquid, referred to as free volume (V_F). The free volume arises from the random motion of molecules and provides the spatial channels required for the migration of gas molecules. In addition, the free volume of a gas molecule is affected by the nature and size of the molecule. The free-volume fraction (FFV) is obtained by calculating the ratio of free volume to total volume. Also, the free-volume fraction (FFV) is temperature-dependent; as the degree of thermal motion of molecules increases, the free-volume fraction within the system also increases, thereby enhancing the diffusion capacity of gases [26,27]. The expression for the free-volume fraction is

$$FFV = \frac{V_F}{(V_O + V_F)} \tag{4}$$

where V_O is the total volume of the system.

Based on the Free-Volume Theory (FVT), we quantitatively analyzed the free-volume characteristics of the synthetic ester and characteristic gas mixture system in the temperature range of 343 K to 473 K using the Connolly surface method and the Atom Volume and Surface tool in Materials Studio software 2020.

As shown in Table 2, as the temperature increases from 343 K to 473 K, the V_O value slightly decreases, while the V_F value significantly increases, leading to an increase in FFV from 0.20–0.25 to 0.31–0.37. This indicates that with the rise in temperature, the thermal motion of the oil molecules intensifies, enlarging the intermolecular gaps and forming more interconnected free-volume regions. As illustrated in Figure 5a–d, the blue regions in the figures represent the free-volume regions, which provide more favorable pathways for gas diffusion. As shown in Figure 5a,d, at 343 K, the distribution of free volume is sparse, and the FFV of H_2 is only 0.20. However, at 473 K, the free-volume region reaches a maximum, and the connectivity of the blue areas is enhanced. As shown in Figure 6a, the FFV of H_2 increases to 0.31, corresponding to an almost 11-fold increase in its diffusion coefficient. This suggests that at higher temperatures, the molecular model of the synthetic ester shows a significant increase in the intermolecular gaps, facilitating the diffusion of H_2 in the insulating oil.

Table 2. Free volume fraction coefficients at four temperatures.

		H_2	CO	C_2H_2	CH_4	CO_2	C_2H_6	C_2H_4
343 K	$V_O/Å^3$	10,350.02	10,485.33	10,533.40	10,529.35	10,505.37	10,666.58	10,625.38
	$V_F/Å^3$	2646.90	3142.39	3464.01	3269.71	3281.19	3356.80	3195.41
	FFV	0.20	0.23	0.25	0.24	0.24	0.24	0.23
393 K	$V_O/Å^3$	10,160.35	10,393.38	10,427.70	10,257.76	10,396.42	10,531.99	10,533.93
	$V_F/Å^3$	3372.90	3366.26	3770.76	4197.00	3944.10	4437.33	3675.62
	FFV	0.25	0.24	0.27	0.29	0.28	0.30	0.26
423 K	$V_O/Å^3$	10,104.14	10,371.54	10,459.69	10,437.51	10,300.77	10,447.81	10,475.45
	$V_F/Å^3$	4037.57	3794.67	3797.13	3965.30	4510.79	4450.45	4269.83
	FFV	0.29	0.27	0.27	0.28	0.30	0.30	0.29
473 K	$V_O/Å^3$	10,004.15	10,161.04	10,069.42	10,126.33	10,319.65	10,322.82	10,165.65
	$V_F/Å^3$	4418.79	4601.97	5883.93	4889.77	4559.66	5706.45	5360.52
	FFV	0.31	0.31	0.37	0.33	0.31	0.36	0.35

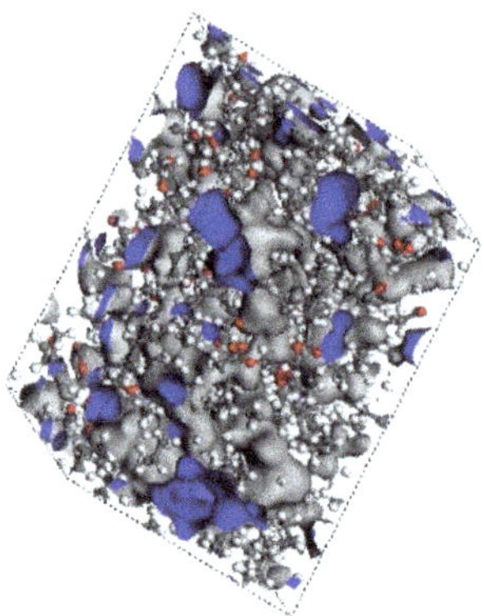

(**a**) The temperature condition of 343 K.

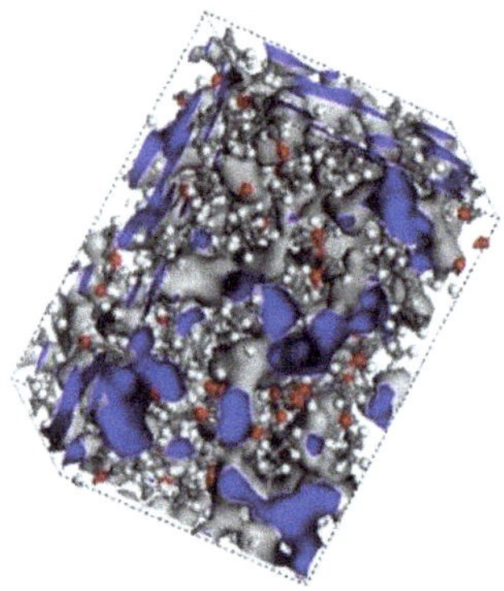

(**b**) The temperature condition of 393 K.

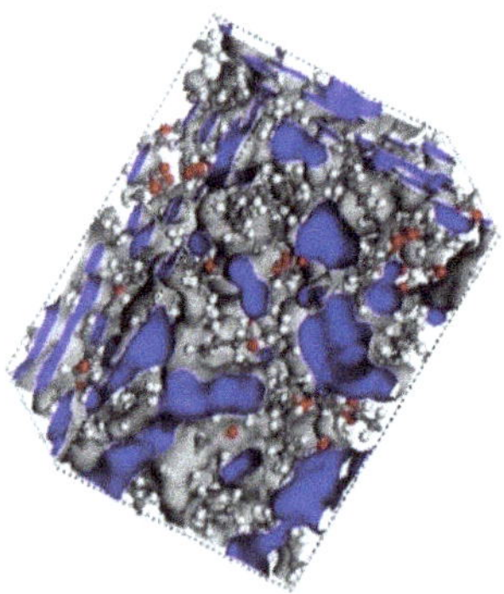

(**c**) The temperature condition of 423 K.

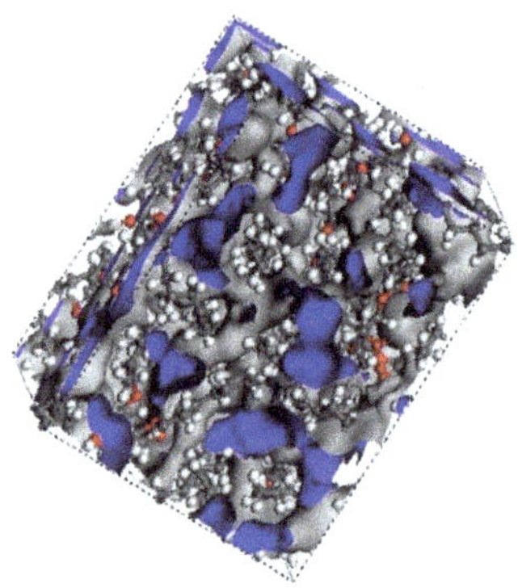

(**d**) The temperature condition of 473 K.

Figure 5. Free-volume modeling of H_2 gas in synthetic ester oils under different temperature conditions. (The blue color in the figure indicates the free volume region in the molecular model).

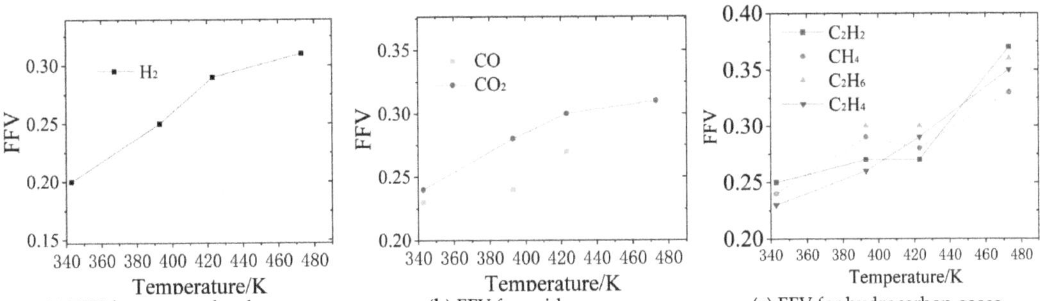

(**a**) FFV for monomolecular gas (**b**) FFV for oxide gases (**c**) FFV for hydrocarbon gases

Figure 6. Trends in FFV values in molecular modeling.

Further analysis of the free-volume variation characteristics of different gases reveals that these characteristics are closely related to the molecular properties of each gas. For small-molecule gases such as H_2 and CO, the FFV is relatively low at low temperatures, being 0.20 and 0.23, respectively, at 343 K. However, at high temperatures, the increase is significant, reaching 0.31 at 473 K, which is consistent with the trend of their diffusion coefficients. In the case of hydrocarbon gases, C_2H_2 exhibits a significantly higher FFV of 0.37 at 473 K due to its high polarity and weak intermolecular interactions, as shown in Figure 6c, so it more easily diffuses into the free-volume regions at higher temperatures in comparison to other hydrocarbons. Although CO_2 also exhibits an increase in FFV to 0.31 at high temperatures, its diffusion coefficient increase is relatively low due to its larger molecular size and stronger intermolecular forces, as shown in Figure 6b.

From the above analysis, it can be concluded that the increase in FFV directly reduces the diffusion resistance of gas molecules. For example, the 12-fold increase in the diffusion coefficient of H_2 corresponds to a 55% increase in its FFV. Furthermore, the variation in the FFV for different gases is in general agreement with the variation in their diffusion coefficients. Therefore, the strong correlation between the free-volume model and the diffusion data validates the applicability of the Free-Volume Theory (FVT) to explaining gas diffusion behavior in synthetic esters. This analysis not only clarifies the microscopic mechanism of temperature's influence on gas diffusion ability but also provides a theoretical basis for revising fault gas analysis methods in transformer insulating oils.

4. Conclusions

In this study, based on molecular models of synthetic esters and characteristic gas molecules constructed using the Materials Studio platform, we employed the PCFF force field and thermodynamic algorithms to quantitatively characterize the diffusion coefficients and free-volume distribution characteristics of gases at various temperatures. Additionally, by integrating molecular dynamics simulations with Free-Volume Theory, the diffusion mechanisms and temperature-dependent behaviors of seven characteristic gases (H_2, CH_4, C_2H_2, C_2H_4, C_2H_6, CO, and CO_2) in synthetic esters were systematically examined within the temperature range of 343–473 K. The results reveal the following.

(1) Under varying temperature conditions, the diffusion coefficients of all the gases studied exhibit a nonlinear increase with an increase in temperature. Hydrogen gas (H_2), due to its extremely low molecular weight and weak intermolecular forces, shows a dramatic increase in the diffusion coefficient, from 33.430×10^{-6} cm^2/s at 343 K to 402.763×10^{-6} cm^2/s at 473 K, reflecting a 12-fold increase. This rate of increase is significantly higher than that observed for the other gases. Among the hydrocarbons, C_2H_2, with its linear molecular structure, demonstrates a notable

advantage in diffusion at elevated temperatures, reaching 99.719×10^{-6} cm^2/s at 473 K.

(2) As the temperature increases, the free-volume fraction (FFV) of each molecular system rises substantially, ranging from 0.20 to 0.37. This increase is primarily due to thermal effects, which expand the intermolecular gaps and diffusion channels, thereby reducing resistance to gas migration. Furthermore, by correlating changes in diffusion coefficients with variations in FFV, it was observed that the 12-fold increase in the diffusion coefficient of H_2 corresponds to a 55% increase in its FFV. This correlation further substantiates the applicability of Free-Volume Theory (FVT) in explaining the diffusion behavior of gases in synthetic esters.

This paper fosters a deeper understanding of the gas diffusion mechanisms in synthetic esters at the molecular scale, offering theoretical insights for the enhancement of fault gas analysis methods in transformer insulating oils. It contributes to improving the efficiency of monitoring the operational status of power equipment, thereby enhancing the safety and stability of power systems.

Author Contributions: Conceptualization, L.G. and H.W.; methodology, W.Q. and W.L.; software, W.Q.; validation, W.L. and J.Z.; formal analysis, J.Z.; investigation, W.Q. and J.Z.; resources, H.W.; data curation, L.G.; writing—original draft preparation, J.Z.; writing—review and editing, L.G.; supervision, H.W. All authors have read and agreed to the published version of the manuscript.

Funding: This paper received no additional funding or sponsorship.

Data Availability Statement: The data that support the findings of this study are available from the corresponding author upon reasonable request.

Conflicts of Interest: Authors Liping Guo and Hongliang Wang were employed by the company Shanghai Electric Transmission and Distribution Experimental Center Co., Ltd. The remaining authors declare that the research was conducted in the absence of any commercial or financial relationships that could be construed as a potential conflict of interest.

References

1. Jovalekic, M.; Vukovic, D.; Tenbohlen, S. Gassing behavior of various alternative insulating liquids under thermal and electrical stress. In Proceedings of the IEEE International Symposium on Electrical Insulation, San Juan, Puerto Rico, 10–13 June 2012.
2. Li, D.; Rao, X.; Zhang, L.; Zhang, Y.; Ma, S.; Chen, L.; Yu, Z. First-Principle Insight into the Ru-Doped PtSe2 Monolayer for Detection of H2 and C2H2 in Transformer Oil. *ACS Omega* **2020**, *5*, 31872–31879. [CrossRef] [PubMed]
3. Liao, Y.; Zhou, Q.; Hou, W.; Li, J.; Zeng, W. Theoretical study of dissolved gas molecules in transformer oil adsorbed on intrinsic and Cr-doped InP3 monolayer. *Appl. Surf. Sci.* **2021**, *561*, 149816. [CrossRef]
4. Wang, J.; Zhang, X.; Liu, L.; Wang, Z. Dissolved gas analysis in transformer oil using Ni-Doped GaN monolayer: A DFT study. *Superlattices Microstruct.* **2021**, *159*, 107055. [CrossRef]
5. Fan, X.; Luo, H.; Liang, F.; Hu, J.; Liu, W.; Li, C.; He, J. Photon Count Technique as a Potential Tool for Insulation Micro-Defect Detection: Principles and Primary Results. *iEnergy* **2023**, *2*, 258–263. [CrossRef]
6. Fan, X.; Niu, S.; Luo, H.; Liang, J.; Liu, F.; Li, W.; Liu, W.; Gao, W.; Huang, Y.; Li, C.; et al. Photon counting technique as a potential tool in micro-defect detection of epoxy insulation pull rod in gas-insulated switchgears. *High Volt.* **2024**, *9*, 267–274. [CrossRef]
7. Jayyid, U.L.; Fadillah, A.; Kurniawan, M.; Suwarno; Rachmawati; Lesmana, A.; Mutia, S.; Anwar, A. Comparative Study of Mineral Oil, Synthetic Ester, and Natural Ester as Liquid Insulation in Transformers. In Proceedings of the 2024 6th International Conference on Power Engineering and Renewable Energy (ICPERE), Bandung, Indonesia, 5–6 November 2024.
8. Lee, S.; Jeong, H.; Park, J.; Seok, B.Y.; Ryu, J.; Bae, C. Research on Dielectric Strength Properties of Synthetic Ester Oil. In Proceedings of the 2024 10th International Conference on Condition Monitoring and Diagnosis (CMD), Gangneung, Republic of Korea, 20–24 October 2024.
9. Guerbas, F.; Adjaout, L.; Abada, A.; Rahal, D. New and Reclamation Transformer Oil Behavior under Accelerated Thermal Aging. In Proceedings of the IEEE International Conference on High Voltage Engineering and Application (ICHVE), Chongqing, China, 10–13 September 2018.

0. Cheng, C.; Fu, M.; Wu, K.; Ma, Y.; Hao, Y.; Chen, C. Aging Effect on Interface Charges between Oil and Oil Immersed Paper. *IEEE Trans. Dielectr. Electr. Insul.* **2021**, *28*, 1579–1587. [CrossRef]
1. Martins, M.A.G.; Gomes, A.R. Comparative study of the thermal degradation of synthetic and natural esters and mineral oil: Effect of oil type in the thermal degradation of insulating kraft paper. *Electr. Insul. Mag. IEEE* **2012**, *28*, 22–28. [CrossRef]
2. Song, H.; Meng, H.; Yuanxin, Y.; Weiguang, Y.; Yupeng, Y.; Yuzhen, L. Influence of Electric Field Distribution on Breakdown Characteristics of Oil-paper Composite Insulation Under DC Voltage. *High Volt. Eng.* **2023**, *49*, 4938–4947.
3. Zhang, Y.; Wang, F.; Li, S.; Zhang, Y.; Li, J.; Huang, Z. Evaluation of Basic Properties and Thermal Oxidation Aging Performance for Synthetic Ester Insulating Oils. In Proceedings of the 2024 IEEE 7th International Electrical and Energy Conference (CIEEC), Harbin, China, 10–12 May 2024.
4. Wu, Y.; Lin, Y.; Wu, P.; Xu, J.; Xie, T.; Gao, C.; Hao, J. Pressure Influence Analysis of Dissolved Gas in Insulation Oil Based on Molecular Simulation and Experimental Comparsion. In Proceedings of the 2021 6th Asia Conference on Power and Electrical Engineering (ACPEE), Chongqing, China, 8–11 April 2021; pp. 756–760.
5. Perrier, C.; Marugan, M.; Beroual, A. DGA comparison between ester and mineral oils. *IEEE Trans. Dielectr. Electr. Insul.* **2012**, *19*, 1609–1614. [CrossRef]
6. Rumpelt, P.; Fritsche, R.; Bakija, B.; Jenau, F. Long-term investigations of partial discharge and gassing behavior of ester-based insulating oils for a heated oil gap. In Proceedings of the International Universities Power Engineering Conference, Heraklion, Greece, 28–31 August 2017.
7. Lashbrook, M.; Al-Amin, H.; Martin, R. Natural Ester and Synthetic Ester Fluids, Applications and Maintenance. In Proceedings of the 2017 10th Jordanian International Electrical and Electronics Engineering Conference (JIEEEC), Amman, Jordan, 16–17 May 2017.
8. Suksagoolpanya, S.; Jongvilaikasem, K.; Jariyanurat, K.; Banthoengjai, T.; Jeenmuang, S.; Pattanadech, N. Dissolved Gas Analysis of Palm Oil Compared with Mineral Oil from Different Types of Breakdown Voltage. In Proceedings of the International Conference on Condition Monitoring and Diagnosis, Bangkok, Thailand, 26–28 October 2020.
9. Suhaimi, N.S.; Ishak, M.T.; Din, M.F.M.; Ariffin, M.M.; Amin, N.A.M.; Hamid, M.H.A. Dissolved Gases Analysis of Rice Bran Oil Under Thermal Fault for Transformer Application. In Proceedings of the 2022 IEEE International Conference on Power and Energy (PECon), Langkawi, Kedah, Malaysia, 5–6 December 2022; pp. 1–6.
20. Hamid, M.H.A.; Ishak, M.T.; Ariffin, M.M.; Katim, N.I.A.; Amin, N.A.M.; Azis, N. Dissolved gas analysis (DGA) of vegetable oils under electrical stress. In Proceedings of the International Conference on High Voltage Engineering and Power Systems, Denpasar, Indonesia, 2–5 October 2017.
21. Williamson, C.; Timoshkin, I.V.; MacGregor, S.J.; Wilson, M.P.; Given, M.J.; Sinclair, M.; Jones, A. Impulsive Breakdown of Mineral Oil and Natural and Synthetic Ester Liquids When Containing Varying Levels of Moisture. *IEEE Trans. Plasma Sci.* **2021**, *49*, 466–475. [CrossRef]
22. Przybylek, P.; Gielniak, J. Analysis of Gas Generated in Mineral Oil, Synthetic Ester, and Natural Ester as a Consequence of Thermal Faults. *IEEE Access* **2019**, *7*, 65040–65051. [CrossRef]
23. Przybylek, P. Drying transformer cellulose insulation by means of synthetic ester. *IEEE Trans. Dielectr. Electr. Insul.* **2017**, *24*, 2643–2648. [CrossRef]
24. Gao, B. Molecular dynamics study on thermal decomposition characteristics of synthetic ester oil. *Chem. Phys. Lett.* **2023**, *813*, 140302. [CrossRef]
25. Tao, J.; Zhan, H.; Luo, C.; Hu, S.; Duan, X.; Liao, M. Diffusion Properties of Gas Molecules in Oil–Paper Insulation System Based on Molecular Dynamics Simulation. *Energies* **2024**, *17*, 3811. [CrossRef]
26. Ye, W.; Hao, J.; Chen, Y.; Zhu, M.; Pan, Z.; Hou, F. Difference Analysis of Gas Molecules Diffusion Behavior in Natural Ester and Mineral Oil Based on Molecular Dynamic Simulation. *Molecules* **2019**, *24*, 4463. [CrossRef] [PubMed]
27. Liu, X.H.; Men, R.J.; Jianng, W.; Li, L.; Wang, W.; Lei, Z. Molecular Dynamics Simulation of the Effects of Nano-SiO_2 Modified by Different Silane Coupling Agents on the Thermal, Mechanical and Electrical Properties of Polypropylene Nanocomposites. *High Volt. Eng.* **2025**. [CrossRef]

MDPI AG
Grosspeteranlage 5
4052 Basel
Switzerland
Tel.: +41 61 683 77 34

Energies Editorial Office
E-mail: energies@mdpi.com
www.mdpi.com/journal/energies

www.ingramcontent.com/pod-product-compliance
Lightning Source LLC
Chambersburg PA
CBHW062043220926
51850CB00010B/2647